RABIC   ASTRONOMY   BANKING   BEE-KEEPING   BIOLOGY
SATION   CALCULUS   CHEMISTRY
MERCIAL CORRESPONDEN   TO
G   CRICKET   DRAWIN   TON
ELECTRICITY IN THE HOUSE   ELO   DERY
NGLISH RENASCENCE TO THE ROMANTIC REVIVAL   ROMANTIC
RYDAY FRENCH   TO EXPRESS YOURSELF   FISHING   TO FLY
BOOK   GARDENING   GAS IN THE HOUSE   GEOGRAPHY OF
ARY   GERMAN GRAMMAR   GERMAN PHRASE BOOK   GOLF
OOD FARM ACCOUNTING   GOOD FARM CROPS   GOOD FARMING
ARMING   GOOD GRASSLAND   GOOD AND HEALTHY ANIMALS
GOOD POULTRY KEEPING   GOOD SHEEP FARMING   GOOD SOIL
HINDUSTANI   HISTORY   ABRAHAM LINCOLN   ALEXANDER THE
CONSTANTINE   COOK   CRANMER   ERASMUS   GLADSTONE AND
ON   PERICLES   PETER THE GREAT   PUSHKIN   RALEIGH   RICHELIEU

· · · AND HE WILL BE YET WISER *Proverbs 9.9*

                                                           LETTER
IN   ANICS
N   ORING
OSO   HYSICS
IBI   PUBLIC
KO   SSIAN
S N   6 AND PURPOSE   SOCCER   SPANISH   SPE   AND
WA   SWEDISH   TEACHING   THINKING   TRIG   METRY
RITH RAILWAYS FOR BOYS   CAMPING FOR BOYS AND GIRLS
GIRLS   MODELMAKING FOR BOYS   NEEDLEWORK FOR GIRLS
AND GIRLS   SAILING AND SMALL BOATS FOR BOYS AND GIRLS
FOR BOYS   ADVERTISING & PUBLICITY   ALGEBRA   AMATEUR
BIOLOGY   BOOK-KEEPING   BRICKWORK   BRINGING UP
CHEMISTRY   CHESS   CHINESE   COMMERCIAL ARITHMETIC
ELLING   TO COMPOSE MUSIC   CONSTRUCTIONAL DETAILS
DUTCH   DUTTON SPEEDWORDS   ECONOMIC GEOGRAPHY
EMBROIDERY   ENGLISH GRAMMAR   LITERARY APPRECIATION
ROMANTIC REVIVAL   VICTORIAN AGE   CONTEMPORARY
HING   TO FLY   FREELANCE WRITING   FRENCH   FRENCH
GEOGRAPHY OF LIVING THINGS   GEOLOGY   GEOMETRY
BOOK   GOLF   GOOD CONTROL OF INSECT PESTS   GOOD
M CROPS   GOOD FARMING   GOOD FARMING BY MACHINE
GOOD AND HEALTHY ANIMALS   GOOD MARKET GARDENING
OOD SHEEP FARMING   GOOD SOIL   GOOD ENGLISH   GREEK
ABRAHAM LINCOLN   ALEXANDER THE GREAT   BOLIVAR   BOTHA
MER   ERASMUS   GLADSTONE AND LIBERALISM   HENRY V   JOAN OF
PUSHKIN   RALEIGH   RICHELIEU   ROBESPIERRE   THOMAS JEFFERSON
E NURSING   HORSE MANAGEMENT   HOUSEHOLD DOCTOR
ALISM   LATIN   LAWN TENNIS   LETTER WRITER   MALAY
NTS   WORKSHOP PRACTICE   MECHANICS   MECHANICAL
ORE GERMAN   MOTHERCRAFT   MOTORING   MOTOR CYCLING
PHYSICAL GEOGRAPHY   PHYSICS   PHYSIOLOGY   PITMAN'S
PSYCHOLOGY   PUBLIC ADMINISTRATION   PUBLIC SPEAKING

THE TEACH YOURSELF BOOKS
EDITED BY LEONARD CUTTS

MECHANICAL ENGINEERING
Volume One

HAND TOOLS
DESCRIPTIONS AND USES

TEACH YOURSELF
MECHANICAL ENGINEERING

A Complete Course in Three Volumes
written and fully illustrated

by

A. E. PEATFIELD
A.M.I.Mech.E., A.M.I.Struct.E.

Volume One
HAND TOOLS: DESCRIPTIONS AND USES

Volume Two
ENGINEERING COMPONENTS
AND MATERIALS

Volume Three
WORKSHOP PRACTICE

Each of these books is a unit in itself,
and may be bought and used separately

TEACH YOURSELF MECHANICAL ENGINEERING

VOLUME ONE

HAND TOOLS
DESCRIPTIONS AND USES

Written and Illustrated by
A. E. PEATFIELD, A.M.I.Mech.E., A.M.I.Struct.E.
Part-time Lecturer Engineering subjects, Willesden Technical College
(Middlesex C.C.)

THE ENGLISH UNIVERSITIES PRESS LTD
102 NEWGATE STREET
LONDON, E.C.I

First published 1950
This impression 1958

Printed in Great Britain for the English Universities Press, Limited,
by Richard Clay and Company, Ltd., Bungay, Suffolk

PREFACE

THIS book is intended primarily for those contemplating an engineering apprenticeship and for student beginners, in order to acquaint them with the various tools to be encountered and their particular uses.

It is hoped the book will also be of use to the layman, who, although possessing no practical mechanical knowledge, may be " mechanically minded " and desirous of carrying out small repair jobs about the house, garden, or garage, or of acquiring some elementary engineering knowledge as a hobby.

Much time and expense can often be saved by using the few tools available to the best advantage, or in their proper manner. Sometimes, however, by the lack of a little practical knowledge, or by attempting to tackle a repair job in a haphazard manner, more serious damage to the object or injury to the operator may result. Knuckles have frequently been badly bruised by a spanner slipping when used in the wrong manner.

This book includes several hints for the inexperienced user of tools. Chapters are devoted to the description of various hand tools, together with their uses and applications. The usual engineering terms are fully explained.

An attempt has been made to submit details in their simplest forms, while maintaining all the essential features.

Many illustrations and diagrams are given in the hope that they will be easily understood. An effort has been made to make them self-explanatory and as clear as possible.

As " Mechanical Engineering " covers such a wide range it will be readily appreciated that it is impossible to describe every tool in a book of this size, but the chief types of those common to everyday use are given.

The two further volumes completing the work deal with "Engineering Components", "Materials Used in Mechanical Construction", "General Workshop Practice"—with introductions to "Pattern-making", "Moulding and Foundry Work", "Forging", "Bench-work and Fitting", etc., "Marking Out", "Machine-shop Work", and "Draughtsmanship".

Finally, the author desires to thank the British Standards Institution for their permission to reproduce certain tables of standard gauges, screw-threads, etc., and Messrs. Broom & Wade Ltd., for supplying information concerning pneumatic riveting tools, also to express his thanks to various friends for reading through the proofs.

A. E. P.

CONTENTS

HAND TOOLS AND THEIR USES

CHAPTER I

HAND TOOLS: DESCRIPTIONS AND USES

As in all professions and trades, the beginner must first
acquaint himself with the various tools he will meet, and
must also learn their different forms and uses.

SPANNERS

Open-jaw Type (Fig. 1)

Let us first of all consider an everyday tool—the

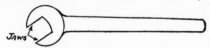

SINGLE ENDED SPANNER   (a)

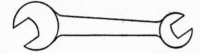

DOUBLE ENDED SPANNER   (b)

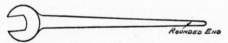

"PODGER" OR "PRONG-ENDED" SPANNER (c)

Fig. 1.

" spanner ". This is, of course, used for the tightening up
or the unscrewing of nuts and bolts, etc. They are usually
made of hard, tough steel in order to stand up to the duties
which they are called upon to perform. The jaws are made
so as to fit comfortably over the flat faces of the nut without
any undue slackness.

9

There are three chief types of open-jaw spanners :—

    (a) The single-ended.
    (b) The double-ended.
    (c) The podger or prong-ended.

These are indicated in Fig. 1.

To use a spanner it should be firmly placed on the nut or bolt-head so that the jaws extend completely over it, as otherwise the spanner will probably slip off when an attempt is made to use it. The smaller jaw-arm should be placed in the direction of the intended " hand-pull ". The larger jaw-arm, having more metal in it, is thus better enabled to take care of the tension force which is exerted, whilst the short jaw-arm tends to be compressed. When a spanner is

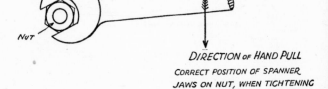

DIRECTION of HAND PULL

CORRECT POSITION OF SPANNER
JAWS ON NUT, WHEN TIGHTENING
UP A RIGHT-HANDED THREADED NUT

Fig. 2.

correctly used in this manner the jaws are less liable to become strained or opened out, and this considerably reduces the possibility of the spanner slipping off, thus avoiding the risk of bruising the operator's hand (see Fig. 2).

The single-ended spanner is used when only one size of nut is to be encountered, and the double-ended type is used in cases where two different sizes of nuts or bolt-heads are to be dealt with. The podger or prong-ended spanner has an open jaw at one end to fit a specified size of nut and a fairly long stem or handle of circular section. This handle gradually tapers, i.e., diminishes in diameter, towards the opposite end from the jaw. This tapered stem, in addition to being used as a handle, may also be used for prising any two plates (see Fig. 3).

It is thus chiefly used for steel-plate work, tank work, etc., in cases where any two bolt holes or rivet holes of plates do not quite coincide. This means that if any two plates which have been previously drilled with rows of holes are to be bolted or riveted together (one on top of the other) it is possible that on placing them into position one or more holes of one plate may be found to be slightly out of register with the corresponding hole of the other plate. In such cases the handle end of a podger spanner may often be used as a drift to prise the plates slightly in order to make the two holes coincide. This procedure, however, is not allowed in boiler work or high-pressure-vessel work, where the two relative holes must be accurately drilled, so as to coincide exactly with each other.

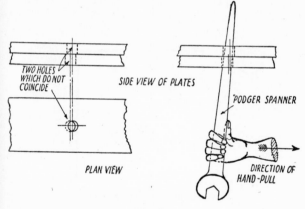

Fig. 3.

Box Spanner

In engineering work several other kinds of spanners are used, the type selected depending chiefly on the accessibility of the bolt-head or nut.

In the construction of various machines, it is sometimes essential to place bolts in positions which are difficult of access by ordinary single- or double-ended spanners. For such cases a special type, known as a box spanner, is used. This type, instead of fitting the nut on two of its faces or sides only, fits it on all its sides, and this permits of greater force being applied without the danger of the spanner opening out. It therefore has an advantage over the jaw-type spanner.

Tubular-type Box Spanner (Fig. 4)

This is perhaps one of the commonest forms of box spanner. As its name implies, it is of steel of tubular section. Both ends of the spanner are, however, hexagonally shaped to fit the intended sizes of nuts. Near each end, immediately above the hexagon formation, a hole is drilled through the tube for accommodating the steel handle or " tommy bar ", as it is called. The hole at one end of the tube is formed at right angles to that at the other end, *i.e.*, the holes are at ninety degrees to each other.

To use this type of spanner, one end is placed over the nut, and the tommy bar is inserted through the hole at the other end. It is sometimes preferable, however, to

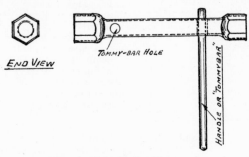

END VIEW

TOMMY-BAR HOLE

HANDLE OR "TOMMYBAR"

Fig. 4.—Tubular Box Spanner.

place the tommy bar through the tube at the end nearer to the nut on which it rests, if this can be done conveniently. The advantage is that none of the pressure applied to the bar is then lost through misalignment of the spanner. It will, perhaps, be realised by the reader that if the bar is inserted through the tube at its top end a certain amount of pressure may be lost by the " torque " or tendency of the tube to twist slightly throughout its length. There may be instances, however, where it is found impossible to insert the bar through the lower end of the tube, in which case the hole at the top end must be used.

Now let it be assumed that the particular nut is situated in a fairly deep rectangular recess of a machine. Let it also be assumed that the length of the recess will easily accommodate the tommy bar, also that it will leave room for reasonable hand movement during the turning operation. The width of the recess may be only half of its length. This being the case, and the tubular box spanner having been placed with its lower end over the nut in such a manner that the lower holes are lengthways or parallel with the " length " of the recess, it is possible for a turn on the bar to be effected.

It might be found that the turn obtained was only an inch or two, say through an arc of about twenty degrees. In that case the hand on the bar will " foul " one side of the recess, thus preventing a greater turning movement. The tommy bar can then be withdrawn and reinserted, but this time in the holes through the top end of the tube. As these holes are at right angles to those at the lower end, a further turn can thus be made. This procedure can be repeated by applying the bar first to one, and then the other end of the tube until the nut is secured.

In certain cases, and in order to overcome this constant changing of the bar positions, a specially long tubular spanner may be employed, by means of which the complete turns can be effected from the top-end position without withdrawing the bar. The tube should be of suitable rigidity, however, to reduce the torque to a minimum, or a certain amount of effective pressure may be lost.

Socket Spanner (Figs. 5 and 5a)

In several respects this type is similar to the box spanner. It fits completely over the whole of the nut, and it also has a separate handle. The socket head is made of one piece of tough steel and is recessed on its lower face to accommodate the nut. On the top of the socket a head or " lug " projects. This is either square or hexagonal according to the form adopted for the handle or " key ". Socket spanners are supplied in sizes which cater for a whole range of nuts. All the socket heads, however, are made to fit one key.

This type of spanner is more rigid in its application than the box type, but it can be used only in places where the nut is either above the face of the work or in a shallow recess which does not exceed the depth of the socket head. It does not damage the corners of the nut, so it has an advantage over the open-jaw type of spanner in this respect.

Some modern types of socket spanners have ratchet mechanisms attached to their handles. When using these it is only necessary to turn the handle through an arc of approximately thirty to forty degrees, and then return it to its starting point. This is frequently more convenient than having to turn the handle through a complete circle or half-circle for each operation. Further, these modern types are so devised that it is possible to operate the ratchet mechan-

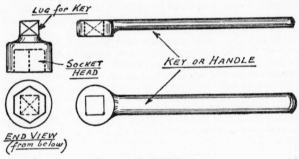

Fig. 5.—Socket Spanner.

ism in the opposite direction. This is accomplished by reversing the ratchet " catch ", which is operated by a small push-knob, causing the ratchet to work in the other direc-

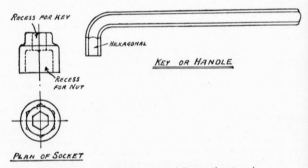

Fig. 5a.—Socket Spanner (alternative type).

tion. It is clearly a great advantage to be able to either screw-up or unscrew a nut without having to remove the socket or handle. The principle of ratchet operation is illustrated in Fig. 80b, but in that case it is applied to a drill.

Other types of socket spanners may have a recess instead of a lug, at their top ends. In these cases the key or handle then has a lug for fitting into the socket recess.

Adjustable Spanner (Fig. 6)

This spanner has jaws, one of which is made integral with the handle, so is therefore " fixed ". The other jaw is attached to a movable slide and is adjustable. By this means several sizes of nuts and bolts can be served. About midway along the handle or " shaft ", at the end of which is the fixed jaw, teeth are formed on the inner edge. Attached to the movable jaw-slide is a small " worm-wheel ". This rotates on a small, steel spindle which is screwed at its rear end, and it is thus fixed horizontally into the slide-jaw frame. By turning the worm-wheel, the sliding jaw can be moved along the teeth of the shaft, as these teeth are " meshed "

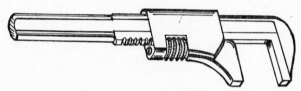

Fig. 6.—Adjustable Spanner.

with those of the worm-wheel. This system gives the adjustment which accommodates the assorted nut sizes.

The chief advantage claimed for this type of spanner is that it is capable of dealing with many sizes of nuts or bolts, thus obviating the necessity of using several different sizes of " fixed-jaw " spanners. This is convenient, especially for maintenance or repair work on the breakdown of a machine, where time is a vital factor.

However, as the jaws are not of the rigid or fixed pattern, they are sometimes liable to be strained or opened out when in use. This may cause them to slip off a nut and also to damage the nut corners. It is therefore essential to ensure that the sliding jaw is screwed up to the nut face as tightly as possible and that it is in compression before applying pressure.

When new, these spanners are quite effective, but after prolonged use they are liable to give trouble. This may occur when wear has taken place on the worm-wheel teeth or the teeth of the " rack ", with which the former teeth engage. The best manufacturers pay close attention to this point and make these components fit as well as possible, so as to avoid any undue slackness or " play ", as it is called, in this mechanism. However, in spite of the maker's precautions, after prolonged use " wear and tear " will result. These parts should be kept clean and occasionally oiled, for if small particles of grit are allowed to enter the mechanism they will quickly cause undue wear.

" King Dick " Type Adjustable Spanner (Fig. 7)

Another form of adjustable spanner is depicted in this diagram; it will be seen that the lower jaw forms

Fig. 7.—" King Dick "
Type Adjustable Spanner.

part of the handle unit with which it is made integral. The top jaw has a shaft or spindle, the lower part of which is screw-threaded. This screwed section has mounted upon it a cylindrical-shaped "nut" which rotates on the screwed spindle, and is operated by turning it between the thumb and finger. The nut is positioned in the guide slot of the handle part. By manipulation of the nut, the spindle upon which it is mounted moves. The top jaw, being part of the spindle, consequently moves, thus opening or closing the jaws. It will be noticed that in this design it is the top jaw which is movable and adjustable. To avoid trouble caused by the entrance of grit the screw mechanism must be kept clean.

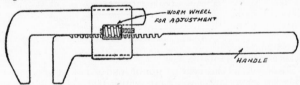

Fig. 7a.—" Monkey-wrench."

Several other types of adjustable spanners are made. All operate on the principle of an adjustable upper or lower jaw and the difference in the design is usually in the mechanism causing the jaw movement.

A very large type of adjustable spanner is sometimes referred to as a " monkey wrench " (see Fig. 7a).

Ring Spanner (Fig. 8)

This is a special design of fixed-end spanner. Each end takes the form of a " ring ", and from this it derives its name. Although of comparatively light construction, it is made of a special type of very tough, hard steel alloy. It is therefore very strong. This spanner usually takes the form of a round or oval-shaped bar, having at each end an " annulus " or ring. The inner face of each ring has teeth on it which fit exactly over the corners of the nut.

Fig. 8.—Ring Spanner.

One or both ends of the spanner are frequently " cranked " in order to cater for nuts or bolts which are in shallow recesses. Although the rings have teeth formations on them, they are arranged to fit snugly over the nut, so do not damage the nut faces or sides. This spanner has therefore an advantage over the open-jaw type, inasmuch as it fits over the whole of a nut in a similar way to a " box " or " socket "-type spanner. It is a modern type, and is usually supplied in sets which cater for a whole range of nut sizes.

In addition to its use in the general mechanical-engineering trade, this spanner is also extensively used in the automobile industry.

Fig. 9.—" Dumb-bell " Spanner.

" Dumb-bell " Spanner (Fig. 9)

This is a special type of double-ended box spanner, and takes its name from its shape, which resembles that of an atheletes's " dumb-bell ". The ends are somewhat spherical, but have their faces flattened. Each face is pierced to form a hexagonal recess of suitable depth to fit effectively over the nut. Each end-piece has four flattened faces with hexagonal perforations of various sizes. This spanner consequently caters for eight different sizes of nuts. Frequently, the two extreme ends of the spanner are also perforated, thus providing for two additional nut sizes.

It is suitable for dealing with nuts which project above the surface of the work but is unsuitable for places where nuts or bolts are in deep recesses. The dumb-bell spanner is a modern type, and is generally made of special steel alloy which is hard and durable. It stands up well to hard use, and is a very handy tool.

General Use of Spanners

From these brief descriptions it will be appreciated that each of the types described has its specific use or application. Although all spanners are primarily intended for screwing or unscrewing nuts and bolts, they may occasionally have to be used for other purposes.

It is fully realised that in the case of emergencies, such as a temporary breakdown of a machine, which might involve loss of production and valuable time, a certain amount of improvisation often has to be done, and in the absence of the specific tool, others have to be used.

It may happen that a particular size of nut has to be removed and that the correct size of spanner is not available. Furthermore, there may not be an adjustable spanner to hand, but there might be an open-jaw type of slightly larger size than the nut concerned. In such cases an old " trick of the trade " is to use a small packing between one side of the nut and one jaw of the spanner. If this is done the " packing-piece " should be placed adjacent to the small jaw-face. Small broken pieces of hack-saw blades can often be used to advantage for this purpose. Further, a small coin has occasionally been used as packing, especially in the case of a motor-car breakdown.

Sometimes, when using spanners, especially the single-ended or double-ended open-jaw types, for securely tightening nuts, it is the practice to place a length of steel pipe over one end of the spanner to obtain a greater leverage for the hand-pull. This method is sometimes used, especially for the tightening up of heavy foundation bolt-nuts, to ensure that they are sufficiently tight to prevent the vibration of the particular machine loosening them. Great caution should be used, however, if such practice has to be resorted to, as if undue pressure is exerted the bolt may be broken by " tearing " or " twisting " across the threaded part. This method should never be adopted with the smaller sized bolts, for the bolts may sometimes break when undue pressure is used on an ordinary spanner.

The length of a spanner is designed to provide sufficient " leverage " for a normal hand-pull or a fair average pressure to be applied. This is why the length of double-ended spanners varies in proportion to the size of their end-jaws. This point should be borne in mind when about to apply a length of pipe to the spanner with a view to increasing the leverage for the hand-pull.

PLIERS

These are used for gripping or holding small objects by manual operation.

Several different types are commonly used in mechanical engineering. Each type has a specific use or application. Pliers are made of tough steel, and some types have their jaws specially hardened in order to adapt them for such uses as wire-cutting, etc.

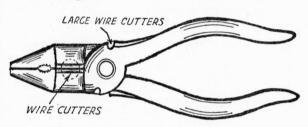

Fig. 10.—Plain or Straight-jaw Type Pliers.

Plain or Straight-jaw Type (Fig. 10)

This type is perhaps the most common in everyday use. It has two jaws, which are slightly tapered towards their ends. The inner faces of both jaws have very small teeth or "serrations" impressed on them, and these enable a firm grip to be obtained on the article which is being held.

Some types have devices attached to their jaws so that they can also be used for wire-cutting and other purposes. In this case recesses are formed about half-way along the jaws on the inner side, and the edge of the recess on each jaw forms a small knife-edge. These are hardened by special heat treatment to render them suitable for the duty which they are called upon to perform. This device is intended for cutting wires of only relatively small diameters. Some types also have two other devices each side of the hinge rivet for the cutting of larger or stronger wires. These devices usually take the form of two small, curved shear blades, which operate by sliding over each other's faces, in a similar way to scissor blades. They are often capable of cutting wires up to approximately one-eighth of an inch in diameter, and are sometimes used for cutting " split-pins ".

Electrician's Pliers

These are usually of the above-mentioned type, but in addition to the details already described their handles are provided with an insulated covering. This usually takes the form of a coating of vulcanite, ebonite, plastic or rubber mixture, or other insulating material. The " insulated " handles thus prevent the user from getting an electric " shock " should the pliers be used on any electrical apparatus carrying an electric current, or if the current has previously been cut-off for some repair work and has been inadvertently switched-on again.

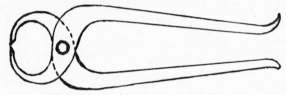

Fig. 11.—Pincer Pliers.

These pliers are insulated sufficiently to withstand several thousand volts, but it is strongly recommended never to attempt repair work, or even to do minor adjustments, on any electrical machinery until the current has definitely been switched " off ", the system " earthed " and every precaution has been taken to ensure that it is kept off until the repair or adjustment has been completed.

Pincer Pliers (Fig. 11)

Although these are not, strictly speaking, " mechanical " engineering tools, they are sometimes used as such. They can often be used to advantage for extracting used split-pins from awkward positions in machinery. However, they are used more frequently for pulling out nails from timbers.

They comprise two parts which are lightly riveted together at the hinge position, as shown in the illustration. The handles are curved to form a suitable hand-grip. The jaws are of circular form, and their thickness tapers from the hinge towards the ends. These ends are usually fairly

sharp and thin to enable them to penetrate into the timber
immediately below the nail head. The end of one handle is
sometimes " fork " shaped so that it can also be used for
" prising " out nails. Pincer pliers are made of fairly
tough steel throughout. Some better-quality types may
have their jaw-ends hardened in order to prolong their life.

Round-nosed Pliers (Fig. 12)

This type of pliers, as the name suggests, has jaws which
are round in section; the jaws are also tapered slightly
towards their extremities. These pliers are used principally
for wire-work—for bending or twisting wires to form curved
shapes or for forming loops, etc. The round formation of
the jaws prevents " kinks " or sharp corners being made in
the wires.

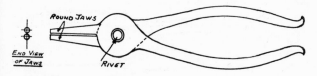

Fig. 12.—Round-nosed Pliers.

Some of these pliers also have wire-cutting devices
similar to those depicted in Fig. 10 positioned near the
hinge-rivet " boss ".

" Quick-grip " Pliers (Fig. 13)

These are so called because they are used chiefly for the
" quick gripping " of square or hexagonal objects, such as
nuts, bolt heads, etc. They can thus often be used instead
of a spanner, and can frequently be used more quickly than
by first having to place a spanner in position on a nut and
then turn it. These actions can be combined in one
operation by using " quick-grip " pliers, but the final tightening up of a nut, to be effectively done, must be by some
kind of spanner.

These pliers are arranged so that when the jaws are in the
normal closed position a small space exists between their
ends. The inner faces of the jaws are tapered towards their
ends, and small teeth are impressed on them to provide for
a sure grip on any object which is being held.

Due to the specially tapered formation of these jaws,
several sizes of nuts can be gripped, by variation of the

opening between the jaws, until their inner faces are opened out beyond the parallel stage. They are thus intended only for use up to and including the parallel stage. It will be realised that if the jaw-opening exceeds the latter stage, the " taper " of the inner faces is then in the wrong direction and thus becomes ineffective. The tendency in this case would be for the jaws to slide off the nut instead of gripping it. These pliers are therefore made in three different sizes in order to obviate this. Each size of pliers, however, is capable of use with a very wide range of nut sizes.

These are very handy pliers, and are useful in cases where nuts require only temporary fixing, so need not be tightened up to the extent of requiring a spanner. They can also be used for obtaining a quick grip on a bolt head in a difficult position and in cases where a special box-type spanner

Fig. 13.—" Quick-grip " Pliers.

would otherwise have to be used. It will be readily appreciated that the primary screwing up of a nut, along the bolt threads, can be done much more quickly with these pliers than with a spanner. This can be accomplished by gripping the nut with the pliers—twisting the wrist to " turn " or rotate it—releasing the grip and re-gripping the nut at a fresh place. The combined operations can often be carried out in far less time than is occupied in positioning a spanner over the nut and performing the same number of turns.

In addition to being used to a great extent in general mechanical engineering, these pliers are extensively used for aircraft-engine work and in the automobile industry.

Gas Pliers (Fig. 14)

In general construction this type of pliers is similar to those previously described, except that the jaws, instead of being of a straight formation, are curved. The inner faces of the jaws also have well-defined saw-edge-type teeth projecting from them.

These pliers are used for gripping and holding cylindrical-shaped objects of relatively small diameters, such as gas- or water-pipes, small rods, etc. They are used extensively on all classes of pipe-work, in connection with the assembly of the fittings—sockets, nipples, unions, etc.

Owing to the efficient grip assured by the teeth of these pliers, they sometimes " bite " into the objects which they are holding. The small marks thus caused can usually be removed by lightly filing the areas affected.

TEETH FOR GRIP· WIRE CUTTER

Fig. 14.—Gas Pliers.

For the holding of larger-diameter pipes and bars, other tools are used. These are dealt with at a later stage.

HAMMERS

Hand Hammer (Fig. 15)

The most common type of hammer used in mechanical engineering is the hand hammer. It consists of a tough cast-steel or forged-steel " head " of about two to two and a half pounds in weight, which is fitted to a wooden " shaft "

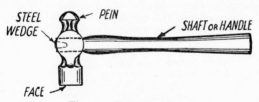

STEEL WEDGE   PEIN   SHAFT or HANDLE

FACE

Fig. 15.—Hand Hammer.

or " handle ". The shaft has a " cleft " or " slot " formed in its thin end. This end is tightly driven through the hole of the head, and into the cleft is driven a steel wedge. Thus any tendency for the " hammer-head " to " fly off " or leave the shaft while it is in use is counteracted by extra pressure on both sides of the cleft. This pressure is transmitted to the sides of the wedge, and retains the head securely in position.

The head has a small cylindrical-shaped " boss ". The flat end of this boss is called the " face " and is used for general " striking " or " hammering " work. The opposite end of the head is " half-ball " shaped, and is called the " pein ". This is used for hand riveting or " burring-over " work.

When in use, either for general work or for riveting, the shaft should be firmly gripped by the right hand at a position about two-thirds of its length from the head. Only for lightly striking any object, or for gently " tapping " a bolt into position, should the shaft be gripped closer to the head.

Sledge Hammer (Fig. 16)

In addition to hand hammers, some hammers of heavier types are used for engineering work. These are called " sledge hammers ". Their heads usually have two equal

Fig. 16.—Sledge Hammer.

and opposite faces, and are commonly of seven, ten, or fourteen pounds in weight; they are attached to shafts of length corresponding to the weight of each head. The head is secured to the shaft in a similar way to that which has previously been described for a hand hammer.

A sledge hammer is intended to be used with its shaft gripped in both hands. This grip should be about three-quarters of the length of the shaft from the head. The shaft-length of a sledge hammer whose head weighs seven pounds is about three feet. When the shaft is gripped at the position described, an effective leverage is obtained, which results in a good striking blow being given. For some classes of work, especially that of hand-forging or blacksmith's work, the blows may be obtained by lifting the hammer into the air for a short distance and bringing the head down smartly on to the object which it is desired to strike.

In an alternative method used where heavier blows are desired, the hammer is " swung " with a complete circular motion. This is frequently more effective, and in certain cases is also less laborious than that described above. When either method is adopted, the user should have his feet well apart, so as to develop an even " poise " or " stance ". This gives an even balance to the " striker ", which is the term applied to the man who uses the sledge hammer.

The best kind of wood for use for hammer-shafts is hickory. Ash, if well seasoned, is sometimes also used.

Sledge hammers are used extensively in heavy engineering for driving large gear-wheel " hubs " or " bosses " into position on shafting during the " keying-up " process. In order to prevent damage to the wheel-bosses, a " buffer " made of lead or copper is frequently held on the face of the boss at the point where the hammer-blows strike.

Special Hammers

In addition to the hand hammer, other steel types are often used for heavier classes of hand-riveting work. These may range from three pounds up to five or six pounds in weight, depending on the nature of the job. Their shafts are often two feet to two and a half feet in length, and are gripped by both hands approximately two thirds of their length from the head. They are used for hot-riveting work of medium size. Often two men stand over the work—side by side or one man facing the other—and each strikes in turn. This procedure is adopted chiefly in order to form the rivet head quickly before it " cools off ".

Except for small jobs, riveting by the use of hand hammers has of recent years been superseded by pneumatic or electrically operated tools and machines. There are, however, certain special cases where the former methods still have to be used.

Hide-faced Hammer (Fig. 17)

The " hide " or " leather-faced " hammer is one which, although its head-centre is made of steel, has the faces formed of stout leather which has been previously treated with glue. The head-centre is usually of hollow, cylindrical form. Into each end is pressed leather in coil formation. The shaft, which is made of wood, passes through the head centre, and is secured in the normal way. In cases where blows are required on finished or semi-finished work the use of this hammer avoids any damage. Hide-faced hammers are made of varying weights up to about five pounds.

Quite heavy blows can be given without any damage being done to the work which is struck. Hide-faced hammers are therefore used as an alternative method to using an ordinary steel hammer in conjunction with a lead or copper " buffer " pad.

This type of hammer is modern and popular. Occasionally, however, for certain classes of work, hammers having copper or solid rubber faces instead of leather are used. The copper-faced type gives quite heavy direct blows, but it is more likely to damage the work slightly. The rubber-faced type, on the other hand, does not damage the work, but owing to the elasticity of the rubber the blows are less

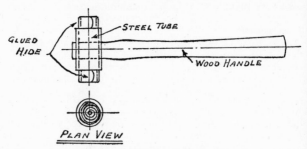

Fig. 17.—Hide-faced Hammer.

effective than copper or leather. In each case it is the nature or class of the work which is the factor deciding which type of hammer is applied to the job.

"COLD" CHISELS

These tools are called " cold " because they are used for work on metals which are in a " cold " state. This is to distinguish them from tools which are used when working on metals in a " hot " state, such as blacksmith's work.

The cold chisel is principally a " fitter's " tool. It is used for " chipping-off " or cutting away small pieces of metal in cases where the desired reduction in size is too much to be easily filed off. In most cases a " cold " chisel is used for roughly reducing the size of an object, and then the final smoothing-off is done by the use of various grades of files.

Cold chisels are made of a good, tough quality of steel,

often called " tool-steel ". They are usually of hexagonal or round cross-section of about one inch in diameter, and eight or nine inches in length. At the lower end a cutting edge is made, thereby forming a blade whose shape varies according to the type of chisel. The cutting edge is heat treated, or " tempered ", thus giving it a hard face. This face is then ground on a grindstone to the required degree of sharpness. The tempering and hardening processes enable the chisel to cut any metals which are not quite so hard as the chisel itself. After prolonged use the cutting edge becomes blunted, and it is then reground.

Flat-type Cold Chisel (Fig. 18)

This type is perhaps the most common of those which are in everyday use in mechanical engineering. As indicated

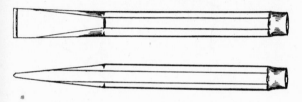

Fig. 18.—Flat-type Cold Chisel.

in the figure, the lower part gradually " tapers " to form a thin, flat end which is ground on both sides to give a fine, hard, cutting edge.

This type of chisel is used for flat " chipping-work ", *i.e.*, on flat surfaces from which it is desired to remove metal. To use the chisel, the work, if small, should be first firmly secured in a " vice " with its surface projecting slightly above the vice-jaw faces. The chisel should be gripped in the user's left hand, at a position approximately one inch from the striking end. A hand hammer of about two or two and a half pounds in weight should be held in the right hand. The hammer-handle should be gripped about two-thirds of its length from the head end. The chisel should be held inclined at approximately twenty-five to thirty degrees from the horizontal, and with the cutting-edge resting on the face of the work (see Fig. 18*a*). A blow from the hand hammer should then be applied, thus causing the chisel point to penetrate the work. The chisel should be given a slight motion in the forward direction, and a further hammer blow applied. This procedure is continued right along the face of the work.

It will be noticed that at each blow of the hammer small pieces of the metal are cut off, thus leaving a somewhat " jagged " or serrated surface. With practice, the beginner will obtain a far smoother surface on his work. It will also be ascertained that some metals are more easily chipped than others—more metal can be cut away or chipped off at one stroke. Cast iron and rolled mild steel, it will be found, vary considerably. With cast iron, it is advisable to endeavour to cut away as little material as possible at each operation if a fairly smooth surface is desired on the work. When chipping mild steel or wrought iron, the beginner

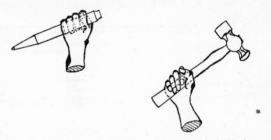

Fig. 18a.—Hand-grips for Hammer and Chisel Work.

may find he develops a tendency to cut too deeply into the material at each hammer-blow. This can be remedied by reducing the angle of inclination of the chisel to the face of the work. It will be found to be of considerable advantage to learn the structure and grain formation of various metals, for most materials will cut far more easily along their grain formation than across it.

In all hand operations it is difficult to acquire the technique solely from written instructions. The beginner is therefore strongly advised, whenever possible, to watch closely an experienced mechanic or fitter at work and to try to copy his actions. After a little patient practice he should become efficient.

When first attempting chipping work the beginner may occasionally miss the chisel-head and inadvertently strike his hand. He should not become discouraged, for this is common to all beginners, and when it happens the tendency

is for him to watch the head of the chisel which he is striking.
This should not be done, however, for he should concentrate
on watching the face of the work which is being chipped.

Normally, chipping work is done at a work-bench, and the
object is held in a vice as described above. In such cases
the operator would be in a standing position. He should
stand half sideways, or half facing the work, and with his
left foot slightly forward, but he should adopt a well-
balanced position or stance Occasionally chipping-
work has to be done to machinery, in kneeling, sitting,
stooping, or even lying-down positions, especially in the
case of large castings. All of these positions are, of course,
more difficult than the normal standing one, until the
operator becomes accustomed to them. The beginner must
first of all get thoroughly accustomed to hammer-and-
chisel work at a bench before attempting it in any other
position. The correct " hammer-swing ", together with the

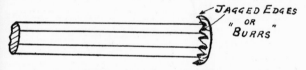

Fig. 18b.—Chisel Striking-head after Prolonged Use.

art of striking the chisel head centrally, will be acquired
after a little practice, and then attention may be given to
chipping in the other positions.

After prolonged use it will be noticed that due to the effect
of constantly striking the chisel head, the latter becomes
slightly flattened and larger than the normal diameter.
There is also a tendency for the head to become burred-
over, and for jagged edges to project all round the striking-
face (see Fig. 18b). When a chisel-head reaches this state
it is dangerous to use it, as small pieces may fly off when it
is struck by the hammer and damage an eye or cut one's
face. These " jagged edges " or " burrs " should therefore
be ground off on a grindstone or emery wheel, and the
chisel-head should be re-formed as shown in Fig. 18. After
much use a chisel requires its point re-sharpening. This is
done by grinding it, preferably on a wet grindstone, which
does not tend to " soften " the blade end. If a chisel is
sharpened on a dry-emery-type wheel which revolves at
high speed, heat is developed owing to the friction of the

surfaces in contact. This takes the " temper " from the end of the blade, thus tending to soften it. If, in the absence of a wet grindstone, the emery wheel has to be resorted to, it is preferable to apply the chisel point only lightly to the wheel, and never to retain it in contact for periods long enough for it to get hot.

When the chisel has been resharpened many times by grinding, the length of its blade becomes shortened or " stunted ". In this case, the blade can be " drawn-out ". This is a " forging " operation which involves heating the whole tool and hammering out its blade to lengthen it. This process, of course, destroys the temper, so that it has to be re-tempered, hardened, and its point reground.

The flat-type cold chisel may sometimes, in the absence of a " hack-saw ", be used for cutting thin sheet metal. The sheet metal is first of all marked along the line of the desired cut, and the flat point of the chisel is placed over the line

SIDE VIEW OF POINT

Fig. 19.—" Cross-cut " Chisel.

before it receives the hammer blow. After each blow the chisel point is moved along the line mark, and at each position hammer blows are applied to it. For this procedure, the sheet metal should be laid on a flat, solid surface. The jagged edges which result from the consecutive chisel-point positions can be " filed up " to give a smooth finish.

" Cross-cut " Chisel (Fig. 19)

A " cross-cut " chisel is one which, instead of having a flat, wide blade end, has a narrow, parallel-sided blade which is ground to a sharp point. It is used for rough preliminary work, prior to the use of the flat type, and in cases where the amount of metal to be cut away justifies the use of both types of chisel.

The cross-cut chisel is also used for chipping-out narrow grooves of square or rectangular section. The width of the chisel blade should thus be the same as that of the desired groove, or if the groove can be reached for later filing up the chisel blade may be slightly narrower than the groove.

HAND TOOLS

" Round-nosed " or Bull-nosed Chisel (Fig. 20)

This type of chisel is one which has a curved, somewhat " boat-shaped ", cutting end. The top side of it is flat. It is used for chipping-out half-round-shaped grooves, including cutting out oil-channel grooves for bearings. For efficient work the cutting edge of a round-nosed chisel must be kept very sharp. The sharpening should preferably be carried out on a wet grindstone.

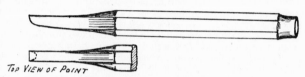

TOP VIEW OF POINT

Fig. 20.—" Round-nosed " or " Bull-nosed " Chisel.

The angle of inclination at which it is held depends on the nature and depth of groove desired. This " slope " or " angle " can be varied to suit the particular job. In some cases, especially in places which may be difficult of access, in order to cut the groove the lower end may have to be bent to form a slight curve.

Side-cutting Chisel (Fig. 21)

The " side-cutting " chisel is a special kind which is used for cutting out slotted holes. A slotted hole is sometimes

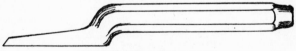

Fig. 21.—Side-cutting Chisel.

made by first drilling a series of round holes which almost touch each other. Then the metal which remains between and at the sides of the holes is chipped away, thus forming a rectangular slot. Finally it is filed up to the desired shape or size.

Nowadays slot-cutting is usually carried out on a machine, but in certain cases, such as those where a " plant " or parts of machinery are already installed, and alterations have to be carried out on the site, the above method often has to be resorted to.

It will be noticed from the diagram (Fig. 21) that the chisel point is sharpened along only one side of its end. This is to enable a slot to be cut evenly by the chisel when it is resting alongside the flat face of the slot concerned. The upper part of the chisel is "cranked" to allow access for a "hand-grip", whilst the cutting blade remains flat.

SCREWDRIVERS (Fig. 22)

This tool, in addition to its chief use—that of screwing or unscrewing screws in wood—is used extensively in mechanical engineering for set-screws of the slotted-head type. It is also used for holding slot-headed bolts to prevent them from turning round whilst their nuts are being screwed up.

The plain type of screwdriver (similar to that illustrated in Fig. 22) consists, essentially, of a steel blade which has a flattened point at the lower end. The handle is usually

Fig. 22.—Screwdriver.

made of wood, but in the case of small screwdrivers which are used for very light work, vulcanite, ebonite, or plastic compositions are used. In size screwdrivers may range between four inches and twenty-four inches in length, the size used depending on the class of work for which it is intended. The blades are made of tough-quality steel, and their points may be "case-hardened" to give them longer life. If a screwdriver's blade point is found to twist at the corners during use, it is probably due to it not being sufficiently hard for the duty it is called upon to perform. The point should not be too sharp, but only sufficiently thin to allow it to penetrate to the base of the slot of the screw or bolt-head.

All-steel Screwdriver (Fig. 23)

A screwdriver which is used extensively in engineering workshops is known as the "all-steel" type. This is made all in one piece. The handle sides are recessed to accommodate pieces of wood which are riveted to the main body. The wood forms a better "grip" than an all-metal handle.

It is also more pleasant, especially in very cold weather, for the user to hold.

This type of screwdriver is very substantial in its construction, and can be tapped with a hand hammer in order to loosen an obstinate set-screw. This procedure cannot be adopted with safety with a wooden-handled screwdriver of the plain type, as a hammer-blow might easily split it.

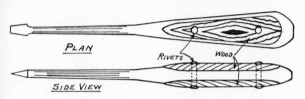

Fig. 23.—All-steel Screwdriver.

Ratchet-type Screwdriver (Fig. 24)

In general outline this screwdriver closely resembles the ordinary plain type, except that it has a " ratchet-device " at the lower shank end of the handle. This arrangement enables the user to reverse the direction of rotation of the blade without removing the latter from the screw-slot. The " reversal " is effected by merely sliding the small

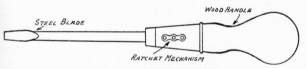

Fig. 24.—Ratchet-type Screwdriver.

knob up or down its guide slot in the handle shank. The blade is not fixed integral with the handle, but in such a manner that it is fitted into the ratchet mechanism. Thus the blade can remain in the slot of the screw which it is desired to turn, and by a mere twist of the operator's wrist the handle is moved backwards and a fresh grip is given by the ratchet ready for the next forward turn. After screwing up any set-screw, it may be unscrewed without removing the screwdriver blade from the slot, by moving the ratchet knob to the reverse position.

B

It will be readily appreciated that the foregoing is a much more simple, and quicker, operation than if the user had to withdraw the screwdriver blade point at each turn in order to obtain a fresh grip ready for his next turn. It is even quicker than if he releases his hand-grip slightly to get a fresh grip in another position on the handle.

The ratchet-type screwdriver is quite modern, and much time can often be saved by its use. Its ratchet mechanism is not complicated, and does not give trouble when it is properly used.

"Pump-type" or "Archimedian" Screwdriver (Fig. 25)

This screwdriver somewhat resembles a cycle-pump in its general driving operation. Its main spindle has double spiral slots. Into these slots fits the movable slide, which is moved up and down the spindle. This action produces

Fig. 25.—" Pump-type " or " Archimedian " Screwdriver.

the twisting movement of the blade. By the aid of the ratchet mechanism the blade can be made to rotate in either a " clockwise " or " anticlockwise " direction. This reversal of direction is effected by the movement of the small knob of the ratchet mechanism to the opposite end of its slot. The operator holds the handle of the screwdriver in one hand, whilst his other hand controls the movable slide ratchet knob, when so desired.

It will be realised that a set-screw can be placed in position in any object, the screwdriver-blade point applied to it, and the set-screw completely screwed up in a matter of seconds. This screwdriver is mostly used for light work, and it is supplied in various sizes up to two feet in length. The spindle slots should be kept free of grit or dust, and the whole mechanism should be occasionally lubricated.

Some types of tool have an adjustable " chuck " or bit-holder in their base, to which various sizes of screwdriver blades can be fitted.

FILES AND FILING WORK

General

Files are made of well-tempered, special hard steel. They are used for accurately reducing the size of any metal object, or for " filing it down " to any desired size, by moving the file over it, and at the same time exerting pressure on it.

A file has a large number of minute teeth formed on its " blade " or " shank ". As these teeth are hard and sharp, they cut out small pieces of the metal when moved over any object. When a very smooth file is used, these " cuttings " or " filings ", as they are called, are only the size of dust. The handle end is sharply pointed, and is called a " tang ", or " prong ". The length of a file is denoted by the length of its " blade ", and the length of the tang is disregarded (except for small files whose handles are made integral).

There are many different kinds of files in common everyday use, the design of each depending on its specific purpose. The name of each kind is usually denoted by its shape, together with the degree of " roughness " or " smoothness " of its tooth formation.

The art of using a file, especially for " dead straight " or perfectly level filing work, requires a considerable amount of skill, which can be acquired only by constant practice and experience. A file should never be used without a proper wooden handle attached to its tang end. If used without one the tang, which is pointed, is liable to penetrate the user's hand.

Types of Files

Some of the chief types which are used in mechanical engineering work are given here, in the respective order of their degree of roughness. They are also given in the order of their use for general work :—

Hand File (Fig. 26)

(a) Rasp, or extra rough cut.
(b) Rough cut.
(c) Middle cut.
(d) Bastard cut.
(e) Second cut.
(f) Smooth cut.
(g) Dead-smooth cut.

Types (a) and (b) are used only for preliminary rough filing work. Types (c), (d), and (e) are used as intermediates, and (*f*) and (*g*) are used for finishing off or for smooth finishing. It is frequently unnecessary to use all those listed, and for most jobs perhaps only (*b*), (*d*), (*e*), and (*f*) would be used. The use of each type naturally depends on the class of work involved. Flat files are used for level filing work or to obtain a flat surface on the work. There are instances, of course, where other types of files will achieve the same result. Flat files are made up to sixteen inches in length and are rectangular in cross-section.

Fig. 26.—Hand File.

Round File (Fig. 27)

Another file in common use is the " round " one. This is cylindrical in shape and has teeth formed diagonally on the surface of its blade. The latter is usually parallel—or of equal diameter—for about three parts of its length, then it gradually diminishes in diameter towards the end, or is said to " taper ". These files are made in many sizes, often up to sixteen inches in length. The diameters are in proportion to the lengths.

Fig. 27.—Round File.

Their purpose is for filing out round or oval-shaped holes in order to enlarge them. The smaller sizes are frequently used for finishing off oil-groove channels in bearings, after they have been chipped out with a special chisel.

Round files are usually made in the following grades, and are used in that order :—

(*a*) Rough cut.
(*b*) Second cut.
(*c*) Smooth cut.
(*d*) Dead-smooth cut.

Half-round File (Fig. 28)

As its name implies, this is semi-circular in cross-section. Its two side edges and the crest of its curve are all parallel for approximately three parts of the length of the blade. The blade then gradually tapers towards its end, where it terminates in a flat shape, but the edges are to some extent " rounded off " or bevelled.

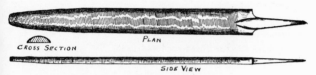

CROSS SECTION   PLAN

SIDE VIEW

Fig. 28.—Half-round File.

This type of file is used for filing hollow surfaces, and sometimes for enlarging holes. Its flat side, however, may be used for " flat " filing work, and its edges for filing in corners of rectangular holes. It is made up to fourteen inches in length, and with its width proportional.

Half-round files are usually supplied in the following grades :—

    (a) Rough cut.
    (b) Second cut.
    (c) Smooth cut.
    (d) Dead-smooth cut.

Square File (Fig. 29)

This file is square in cross-section, and has parallel edges for approximately three-quarters of its length. The

CROSS SECTION

Fig. 29.—Square File.

remainder gradually tapers, but retains its square section.

A square file is used chiefly for filing out rectangular holes or slots. In certain cases, however, it may also be used for filing narrow, flat surfaces.

Some types of square file are supplied with " single-cut " teeth, *i.e.*, the teeth are cut crossways but in one direction only, whilst others may have " double-cut " teeth. The double-cut types have a second row of teeth which crosses the first one. The length of the file may vary from six inches to sixteen inches, and the size of the cross-section varies in proportion to the length.

These files are mostly supplied in the following grades, but other grades are commonly obtainable :—

    (*a*) Rough cut.
    (*b*) Second cut.
    (*c*) Smooth cut.

Three-square File (or Three-cornered Type) (Fig. 30)

This type, as the name infers, is " three-cornered " in cross-section. The width gradually diminishes towards the blade end.

It is often used for similar purposes to the square file, but

CROSS SECTION

Fig. 30.—Three-square File (or Three-cornered Type).

in addition, it is useful for penetrating into the corners of rectangular holes and slots. The " three-square " file, owing to its shape, which results in the formation of very keen cutting edges, lends itself to, and is frequently used for, forming a fine mark on any metal object, to define the exact position before commencing to saw.

These files are usually supplied in lengths ranging from four inches to approximately twelve or fourteen inches. The blade widths vary in proportion to their lengths.

The most common grades are probably the following :—

    (*a*) Rough cut.
    (*b*) Second cut.
    (*c*) Smooth cut.
    (*d*) Dead-smooth cut.

Special Files

In addition to the foregoing common types of file, several " special " or not so common kinds are made.

Amongst these are the following :—

Cotter File (Fig. 31)

This type is similar in general form and shape to the flat file, except that the edges of the rectangular cross-section are " rounded ". The teeth are usually double-cut. It is used mainly for filing out slots for cotters; hence its name.

The cotter file is generally made in smaller sizes which

Fig. 31.—Cotter File.

range from four inches to approximately eight or ten inches in length. It is also often used for plain, flat-surface filing of small classes of work.

It is usually made only in the following two grades :—

(a) Second cut.
(b) Smooth cut.

Knife-edge File (Fig. 32)

This type, as the name infers, is similar in shape to an old-fashioned bread or carving knife. Both its edges are curved, and it has flat sides which form a " wedge " shape in cross-section. The sides generally have their teeth cut

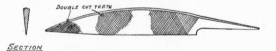

Fig. 32.—Knife-edge File.

crossways, or double-cut, but the edge of the wedge has single-cut teeth, i.e., cut in one direction only. It is used mainly for filing out acute angular-shaped recesses or corners.

This file is supplied in the following three grades :—

(a) Rough cut.
(b) Second cut.
(c) Smooth cut.

Cross-cut File (Fig. 33)

This type usually has its teeth cut " single " fashion. The handle is often made integral instead of a separate wooden handle being used. The sides taper gradually. It is chiefly used for resharpening the teeth of agricultural machinery, such as grass-mowing machines, reapers and harvesters, etc., during the harvesting of crops.

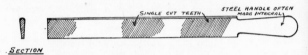

Fig. 33.—Cross-cut File.

A common size used for these purposes is the " ten " or " twelve inch ".

The wide edge of this file is rounded, and the opposite edge has square-shaped corners.

It is chiefly supplied in the following grades :—

    (a) Second cut.
    (b) Smooth cut.
    (c) Dead-smooth cut.

Fig. 34.—Needle Files.

Note.—Types (a) and (b) shown Single Cut. Types (c) and (d) shown Double Cut.

Needle File (Fig. 34)

These are relatively very small, and range between three inches and seven or eight inches in length.

They are mainly supplied in round, half-round, square, or three-square forms.

Needle files are used for delicate or very light classes of

work, and are therefore chiefly made only in the smoother grades.

They also have small, round handles formed integral with the blades.

These files being of a hard and brittle nature are themselves delicate, and should be used with care, for they easily break if undue pressure is exerted on them.

They are used for instrument work and all classes of light jobs, such as cutting out the formation of the letters or figures of thin metal " stencils ".

General Filing Work

Having introduced to the reader some of the chief types of files, the process of " filing " or the " art " of using the various files may now be considered.

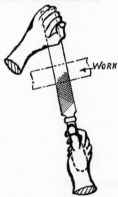

Fig. 35.—Filing (General Work).

The beginner should learn the art of filing by practising on any odd bits of scrap metal; preferably pieces of mild steel or cast iron of approximately three or four inches in size and rectangular in section. The object should be securely fixed in a vice with the " face " of the work which is to be filed raised just conveniently above the vice-jaw edges. He should commence with a partly used flat file, as a new one cuts too keenly. The file handle should be grasped in the right hand, and the left hand should grip the file blade end with the fingers pointing downwards (see Fig. 35). The file should be placed firmly on the face of the work with one side of its blade resting flat on it. The blade should be placed in such a manner that three-quarters of its length is near the user and the left-hand grip of the end is immediately beyond the work.

The beginner should next gently push the file along the surface of the work, whilst at the same time he should endeavour to use even downward pressure applied by both hands. He should also try to keep the forward " push " stroke as horizontal as possible.

This stroke should be continued until the " ferrule " of the handle is approaching the work. At this stage the push stroke should cease, the pressure on the file be released, and the blade drawn back lightly over the work to the starting position. On the backward stroke it can be lifted bodily off the work, for the file teeth are formed solely for cutting during the forward or push stroke. The stroke should then be repeated, while all the time an attempt should be made to retain the file in a horizontal plane.

After a few strokes the beginner should test his work. This is done by placing a straight edge over the surface lengthways or in the direction of his filing. In all probability this will prove that he has developed a slight curve on the surface, and it will be of a convex nature. This is because, although every effort was made to file in a horizontal direction or plane, he has failed to do so. It is probably due to the fact that although the right hand has been kept in a rigid plane with the elbow, the latter moves from the shoulder in an arc or semicircular motion. During the next attempt, the right hand should be raised slightly so as to lower the elbow, and the motion between the shoulder and the hand should be of a reciprocating nature, using the elbow joint as a " pivot " or fulcrum.

The beginner should adopt a stance with his left foot well forward ; with his right foot turned slightly outwards, and his weight evenly balanced on both feet. He should also stand half sideways towards the right, leaning slightly forward, especially for the push or cutting stroke. The stance should be similar to that adopted when using a hammer and cold chisel for chipping. It is just as important to adopt the correct stance for filing work as it is for a boxer, or in cricket, for the batsman to do so.

In order to reduce a curvature which may have developed on the work, the beginner should next reduce the length of his forward stroke and concentrate his efforts on the top of the curve, neglecting to file the edges. He might also try filing the work " across " instead of lengthways, *i.e.*, at right angles to his first efforts. The straight edge should frequently be tried to ascertain the amount of improvement which has taken place.

Some beginners may find it helpful to " tilt " the file slightly or incline it, and, instead of filing precisely forwards and backwards in a straight line, try to file slightly towards the right, or in an outward direction. Incidentally, when a file is tilted, the trailing edge of a partly used file will file off more metal than when it is used straight.

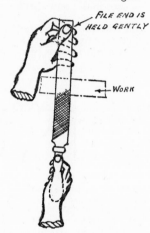

Fig. 36.—Light Filing Work.

The art of filing " dead level " requires a considerable amount of skill which can be acquired only by constant practice, so the beginner should not be discouraged if his first efforts are unsuccessful.

Some metals, especially those of a softer nature, are easier to file than others. The beginner is strongly advised to acquaint himself with the various metals and alloys, especially their " grain " formation, which will materially assist him in his career.

Occasionally when dealing with cast iron, " hard spots " may be encountered in isolated places. These spots will spoil a new file and wherever possible, they should be " ground down " by applying an emery stone to them.

For finishing off work or light filing, especially when using files of the smoother types, the file should be held firmly in the right hand, but gently by the left hand, with the fingers resting below the blade end and the thumb on the top side of the blade (see Fig. 36).

Another method for finishing off work or for erasing deep scratches or deep " rough-cut " marks from work is called " draw-filing ". This is the term used when a file, instead of being held with its blade lengthways over the work, is used " crosswise " or across the object to be filed (see Fig. 37). When used in this way, the file teeth do not cut so " harshly ".

With constant or prolonged use, file teeth become " choked " or " clogged up ", due to the " filings ", *i.e.*, the minute bits of metal which have been cut off, entering the

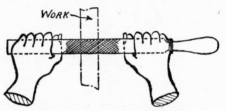

Fig. 37.—" Draw-filing " Work.

recesses between the teeth. The file should therefore be cleaned periodically. This is effected by first, gently tapping it or knocking it lightly on a wood block in order to " shake out " as many of the filings as possible. Next, the file blade should be brushed crossways, using a special wire brush which is provided for this purpose. The brush is composed of stiff wire bristles which are woven on to strong canvas. The canvas is nailed to a suitable wood block which thus forms the handle (see Fig. 38).

When the beginner has had some experience of using a flat file, attempts may be made with other types.

In addition to filing work being done in a standing position it has to be done occasionally in sitting, kneeling, or even lying-down positions, according to the nature of the job. Before attempting any other position, the beginner must first become thoroughly proficient in the straight, level filing of work in the standing position.

A new file should never be used on the outer faces of " castings " which have been in contact with moulding sand, which is used in the operation of casting. A new iron casting, even after being " fettled ", *i.e.*, cleaned up, often contains small particles of sand which rapidly blunt the teeth, or take the edges off a new file. It is therefore advisable always to commence work on the outer faces of a casting with an old or partly used file.

No attempt should be made to file hardened-steel objects, such as cold chisels, drills, shears, etc., manganese steel, or special hard steel alloys. These are not intended to be filed, and are best dealt with by grinding them on the various Carborundum (trade name) emery stones or wheels especially made for that purpose.

When files become badly worn after prolonged use, or the normal " wire brushing " does not improve them, they can

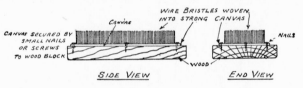

Fig. 38.—File Cleaning Wire-brush.

be returned to the makers for " re-conditioning " or " re-cutting ". The file is then " softened " and the teeth re-formed or re-cut. It is then hardened again and called a " re-cut ". Re-cut files, however, are rarely quite so good as new ones.

In order to produce an " extra fine " finish on filed work, various grades of emery cloth are used. These range from coarse grades down to very smooth or fine types. They are formed of canvas to which is applied the various grades of emery powder. The powder is sprinkled on to one side of the canvas, the latter having previously been given a coating of some sticky substance of a " gluey " or " gummy " nature, to which the emery powder or particles adhere.

After completing the filing by the use of " smooth " or " dead-smooth " files, a piece of emery cloth of the desired grade is " lapped " several times around an old file blade or a piece of flat wood, and used in a similar manner to ordinary filing.

The use of emery cloth in the fashion of draw-filing is often popular. If the finer grades of emery cloth are used, the result will be a very smooth or semi-polished surface. The use of fine grades of emery cloth also assists in obliterating fairly deep file-cut marks or scratches from the work. In order to reduce the " harshness " of new or unused emery cloth before applying it to the work, it may be lightly smeared with thin oil. This procedure, in addition to " killing " the harshness, also assists in giving a polished or smooth finish to the work.

For a " superfine " finish of a " dull " nature, however, very fine grades of emery-powder or crocus-powder, " crocus martis "—(yellow oxide of iron), can be mixed with thin oil and applied with a piece of smooth rag or cloth. The work can then be completely finished off, or highly polished, by using any of the domestic or household types of metal polish in the usual way. Brass or copper articles lend themselves very well to the foregoing processes, and give very satisfactory results.

For repetition work in large, modern factories the final polishing up is executed by various grades of " buffing " wheels. The wheels, which consist of discs of cloth to which some type of metal polish has been applied, are mounted on spindles and rotated at very high speeds. The polishing is thus effected by mechanical methods instead of manual processes.

The beginner should note that the softer metals and alloys—brass, copper, lead, solder and aluminium, etc.—rapidly tend to clog up the teeth of files, and emery cloths, which should therefore be frequently shaken or cleaned during use. It is also worthy of note that after prolonged use of the coarser grades of emery cloths, their cutting or grinding power is considerably reduced, or the harshness is diminished. These partly used cloths should not be too readily discarded, for they may often be found of use in the absence of new emery cloth of the finer grades. They are especially useful and " fine " after a drop of oil has been added to the partly used cloth, particularly if it has not previously been oiled.

SCRAPERS (Figs. 39 and 40)

These tools are made from very hard steel. They have sharp edges and are used for " scraping " objects, the sharp edges cutting away very small pieces from the metal.

Scrapers in general shape and size resemble files; in

fact, many engineering workshops make their own scrapers from old, discarded files from which the teeth and edges have been ground, in order to make them sharp. The shapes of the chief types of scrapers in common use are the flat, the half-round, and the three-square. Other types are occasionally used for special jobs.

The object of scraping any article is to ensure a more accurate fit. Even where parts have been carefully filed up, or if they have been " turned " and " machined " in a lathe or boring machine, the parts may not fit accurately enough for some classes of work.

Scraping is often resorted to for either flat or hollow surfaces. In most cases a flat type of scraper is used for flat-surface work. The half-round scraper is chiefly used on hollow-surface work.

Flat scraping work is usually carried out in conjunction with some predetermined flat surface, which is specially

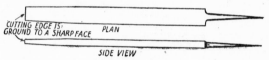

Fig. 39.—Flat Scraper.

made for the purpose. The special surface is known as a " surface-plate " (see Fig. 41). However, when work is not carried out in conjunction with a surface-plate, two parts may be scraped so that the surfaces of the parts " fit " or " bed-down " to each other. The latter method is usually adopted for scraping the "steps" or " liners " of bearings, to ensure an accurate fit between the shaft and the bearing.

For flat scraping, and when the end of the scraper forms the cutting edge, the handle should be held firmly in the right hand, as indicated in Fig. 36 (for filing). The left hand should grip the scraper blade about half-way down its length. The right hand should make the forward push-stroke, and the downward pressure should be applied by the left hand.

Let us next consider the method which is employed in order to scrape the flat surface of any small machine part or " work-piece ". A surface-plate having been made available, some method must be used to show which places on the surface of the work-piece exactly fit or bed-down to the surface-plate, and also to show those parts which are hollow or do not contact accurately the face of the surface-plate.

This procedure is usually effected by making up a mixture of fine red-lead powder and thin oil to form a thin, coloured paste. A coating of this mixture is applied to the surface-plate face in such a manner as to produce a very thin film all over the face. The face of the work-piece, having been cleaned with a piece of soft, dry rag or cotton-waste, is then applied to the surface-plate face. The work-piece is then pressed lightly but firmly to the surface-plate and rubbed against it once or twice. The two surfaces are then separated and that of the work-piece inspected.

It will be found that any " proud " or " high spots " on the work-piece face will have marks of the red-lead mixture on them, whereas any hollow or concave spots will still be clean, and will therefore show no red-lead marks.

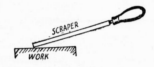

Fig. 39a.—Scraping a Flat Surface.

The work-piece may then be placed in a vice, face upwards, and the " high-spots " carefully scraped down (see Fig. 39a).

The work-piece should then be detached from the vice and thoroughly cleaned.

Now, the next step is *not* to apply it to the surface-plate, as this will bear or carry " smear-marks " which will have originated from the work-piece's previous contact with it. These smear-marks on the surface-plate must therefore be cleaned off, and a fresh red-lead film evenly distributed over the whole of its surface exactly as was done for the first test. Only after the surface-plate has been so treated should the work-piece face be placed in contact with it again. The latter should be gently rubbed over the surface-plate face once more, and again it should be inspected.

These processes are repeated, and each time any high spots are detected, they should be carefully scraped down until the whole surface exactly coincides with that of the surface-plate.

The beginner should note that the art of scraping efficiently requires a considerable amount of skill. No attempt should therefore be made at scraping until he has had considerable filing experience, as the arts of filing and scraping are allied to each other. The more experience which the beginner acquires with filing work, the better enabled he will be to take up scraping in due course.

Having described flat scraping, consideration may now be given to the scraping of curved surfaces. This type of scraping applies particularly to the inner faces of bearing linings of the brass or gunmetal types. In engineering such bearings are called " plummer-blocks " or " pedestals ", and their liners are called " brasses " or gunmetal " steps ". The latter are of hollow, cylindrical shape and are made in two halves, one of which fits into the pedestal base, and the other half fits into the pedestal cap or top-piece. Bearings or pedestals are the components in

TYPICAL SECTIONS
OF VARIOUS SCRAPERS BLADES

Fig. 40.—Curved Scraper.

which steel spindles or shafts are mounted, and in which the shafts rotate.

In order to acquaint the beginner with the necessity for accurately scraping these steps for pedestals, it might with advantage be explained why this procedure is often so essential. It is also hoped that such explanations will better enable the reader to understand the whole procedure which is involved, and which leads up to the necessity of scraping in order to obtain an accurate fit.

A bearing is designed to carry a shaft, whose load is transmitted to the former. This load is intended to be evenly distributed over the area of the lower " half-step ". It therefore follows that the surface area of the shaft must contact the whole surface area of the half-step. Now if this half-step has " places " on its surface which are slightly convex, and other places or spots which are concave, the shaft load will be transmitted only to the convex spots, or places, instead of over the whole area. This considerably increases the stress on those parts, so that the bearing is likely to fail. In order to prevent this occurring, scraping is resorted to during the assembly of the parts of the bearing.

The procedure is as follows :—

The whole bearing is assembled with its lower half-step in position, but with its " cap " off. The shaft is thoroughly cleaned up, and a thin coating of red-lead mixture applied evenly to it. The shaft is next laid gently into the bearing half-step in the position which it will occupy when in working order.

The top half-step together with the bearing cap are then fitted and the shaft gently rotated. The cap and top half-step are then dismantled, and the shaft extracted from the bearing. On examination of the lower half-step it will be observed that impressions have been left on the " proud " or convex spots, where the shaft came into contact with them. Having thus located or detected these spots, the half-step can be removed, placed in a vice, and the high spots scraped.

For this operation, however, as the half-step has a curved inner surface, a half-round-type scraper would be used.

Fig. 40a.—Scraping a Curved Surface.

The scraper handle should be held by the right hand in a similar manner to that which has previously been described for holding a flat scraper. The left hand may either grip the extreme end of the scraper blade or in some cases it may be found preferable to grip the handle end, thus holding the scraper at the latter end by both hands. In all probability the type of scraper which would be used would be the curved half-round, which can be held by the left hand in either of the two positions which have been mentioned (see Fig. 40a).

For finally finishing off the scraping a more delicate touch or a better control is often obtained by holding the blade end in the left hand, giving an even and regular stroke for the cutting edge of the scraper, which in this case is the side edge. Should the half-step be only of short length, probably a three-square scraper could be used instead of a half-round type.

After a perfect " bed " has been obtained for the lower half-step of the bearing, attention may be given to the top half-step. It is not so essential for the latter to have as good a " bed " as the lower one, as it is the lower one which takes most of the load, provided the power applied to the

shaft is of a downward nature, such as may be transmitted
from the belt drive of a motor which is situated at floor
level and below the level of the shaft concerned.

Surface-plate (Fig. 41)

This is an instrument or tool which has on one side of it a
perfectly flat surface. It is used as a " basis " or template
on which work is tried during filing or scraping operations in
order to ascertain whether or not the work is perfectly flat.

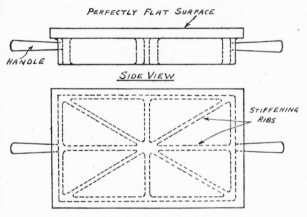

Fig. 41.—Surface-plate.

It is usually rectangular in shape and made of a good-
quality, close-grained cast iron. On its base a surface-
plate has several stiffening ribs. The latter give support to
the " face " and prevent any tendency for it to " sag " or
" warp ".

The size of a surface-plate varies according to the size of
work for which it is likely to be used. It may range from
about twelve inches long by nine inches wide up to about
two feet three inches long, by one foot nine inches wide,
with a corresponding thickness.

Surface-plates are usually made three at a time, in order
that the surface of each can be made interchangeable, and
to ensure that all three are " dead " level. If only two

plates were made, one of them might be slightly convex on its finished surface, while the other, or second plate, might have a concave surface, yet each would fit the other perfectly, and neither of them would be dead level. By making a third plate and fitting it accurately to the other two, all three surfaces are made perfectly level. When making surface-plates they are first carefully cast, and their surfaces then machined. The latter are then very carefully scraped to take out all the tool marks and rough places. One surface-plate is cleaned up and given a coat of red-lead-and-oil mixture, and one of the other two plates is carefully " bedded-down " to it.

The third plate is similarly dealt with, and is bedded-down effectively to each of the other two.

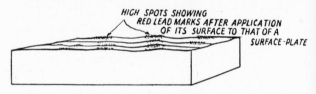

Fig. 42.—" High Spots " on Work Marked by Red-lead after Application to Surface-plate.

A surface-plate usually has a lifting handle at each end to facilitate placing it in position on a bench ready for use.

It is advisable to always keep the flat surface of a surface-plate well oiled or greased in order to prevent it from rusting when not in use, and its surface should be well protected from knocks which may mark or dent it.

A surface-plate may also be used in conjunction with a " set-square " to ascertain if the sides and ends of any rectangular or cylindrical object are perfectly square with the base.

For work of too large a nature for the use of a surface-plate, however, other means would be employed—probably a large flat table which is called a " marking-off " table. This will be dealt with at a later stage.

A mixture of red lead and oil, which must be of a thin nature, is evenly distributed over the face of the surface-plate. The work-piece to be tried is then applied to this surface and is lightly pressed against it. It is also usual to rub the work-piece along the surface-plate slightly.

Caution should be used in doing this, however, in order to avoid the mixture being " thickened up " in certain places from whence it is liable to penetrate between the two surfaces and give a misleading result. On parting the two surfaces, the high spots on the work-piece will be found to show the red-lead marks at the exact contact positions.

The low spots or low places will therefore be quite clean. The former can be either filed or scraped down, and further " tests " made by repeating the procedure until the desired result is attained (see Fig. 42).

Between each subsequent test, however, both the surface of the work and that of the surface-plate must be thoroughly cleaned and a fresh thin film of the red-lead mixture evenly applied to the surface-plate. Great care should also be taken to ensure that no particles of grit are on the coated surface-plate before the work-piece is applied to it.

This is important, because the oil of the red-lead mixture has an affinity for attracting or collecting grit particles, which may be blown about in the atmosphere of a work-shop.

HACK-SAW (Fig. 43)

This is the engineer's hand-saw, and is to him just what a hand-saw is to a carpenter or joiner. A hack-saw consists of a metal framework with a wooden handle at one end, and a cutting blade fitted to the frame. Most hack-saws have frames which are adjustable to accommodate various lengths of saw-blades. This is because some classes of work demand longer blades than others.

Projecting from the lower ends of the saw-frame are rectangular-shaped slides. Each slide has a small peg attached to it for which a hole is provided in the saw-blade. One of these slides has the handle attached to it, and the other slide has a screw-threaded end on which is mounted a wing-nut.

By " screwing up " this wing-nut after the saw-blade is in position, the former tightens up and thus " tensions " the saw-blade.

It must be kept tight, for the saw-blade is of a thin nature, and being made of hard steel is also very brittle. It will not bend without the risk of breaking, so it must be retained in a tight and straight-line position.

The saw-blade has small cutting teeth on its lower edge. The type of the teeth varies according to the nature of work for which they are intended. Some blades have " fine " or very small teeth, whilst other types have teeth of a larger or coarser character.

The length of hack-saw blades varies from eight up to twelve or fourteen inches. These blades are made of special hard steel which has been well tempered or heat treated in order to enable them to cut any metal or alloy which is of a lesser degree of hardness.

To use a hack-saw, the handle should be firmly gripped in the right hand, whilst the left hand should be positioned around the opposite end of the frame, but at the top, just at the point where it is usually bent to form the vertical stay-support which carries the blade end. To start sawing, it is usual to make a small " cut " with the corner of a file at the exact desired position on the object which is to be sawn. This procedure ensures that when the

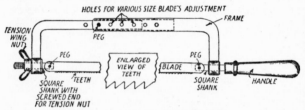

HOLES FOR VARIOUS SIZE BLADE'S ADJUSTMENT — FRAME

TENSION WING NUT

PEG

PEG

PEG

ENLARGED VIEW OF TEETH

BLADE

SQUARE SHANK WITH SCREWED END FOR TENSION NUT

TEETH

SQUARE SHANK

HANDLE

Fig. 43.—Hack-saw (old type).

hack-saw blade is applied to the work it does not " wander " or " stray away " from the position. A three-square or a half-round file edge should preferably be used for making this " starting mark ".

The front end of the hack-saw, *i.e.*, that which is held by the user's left hand, should point slightly downwards, or in other words, the right hand should be in the higher position of the two. The stance should be somewhat similar to that adopted for filing-work.

Most hack-saw blades are designed for the forward stroke to be the cutting stroke. A certain amount of cutting or sawing, however, is also effected by the backward stroke. When commencing to saw, very little pressure should be applied, because sometimes, especially with new blades, they " bite " too keenly and difficulty is occasionally experienced until two or three strokes have been made.

The beginner, when starting to saw, may experience a

little difficulty through the blade developing a tendency to "bind" in the work. He may be tempted, in order to "free" the blade, to "tilt" it slightly sideways. This is fatal, for the blade will easily "snap" if so treated.

He should, therefore, always endeavour to saw in a perfectly straight line, and try to avoid any side movement or "whip". If, however, the blade should bind or become fixed and locked in position in the work, it should be gently lifted out of the "cutting" with a vertical motion, thus avoiding breaking off the small teeth from the blade.

He should then commence again, by placing the extreme end of the saw-blade that is farthest away from him, at the commencement of the cutting and gently and lightly give a forward cutting stroke. Naturally, some metals are more easily sawn than others. Brass is easier to saw if the fine-tooth type of blade is used.

Should a blade be broken during the cutting of a fairly wide work-piece a new blade must be fitted. After fitting it and before commencing to saw, the work-piece should be carefully inspected in order to see that no small pieces or broken-off teeth remain in the slot, for these will cause havoc to a new blade.

The beginner should not try to saw too quickly, for heat is generated by this action, and if continued it will cause softening of the teeth by taking away the "temper" from them.

It is also risky to saw too quickly, for should the saw-blade suddenly bind in the slot, the blade will probably be broken before the user can "pull up" or stop sawing.

When sawing with a hack-saw, the work-piece, if small enough, is usually secured in a vice. Some objects, on account of their shape, might be found difficult of access by the blade, after it has penetrated a certain depth into the slot or cutting. This depth is governed by the distance between the saw-blade and the top of the saw-framework. Owing to the slide-pieces which secure the blade at the frame ends being of a square section, in such cases the wing nut can be unscrewed, the blade and slide-pieces withdrawn and replaced at an angle of ninety degrees from that of their former position. By this arrangement the blade is still in a vertical position in its cross-section, whilst the frame is now horizontal. Sawing can therefore proceed unrestricted by the frame, and the blade can now go right through a thick work-piece.

For sawing with this arrangement, however, the left hand should grip the saw-frame at the wing-nut position.

Tubular-frame Hack-saw (Fig. 44)

Some modern types of hack-saw are now made with a tubular-steel framework. These are frequently nickel or silver-plated, chiefly for appearance's sake. Some types also have " Bakelite " (trade name) or plastic handles of the " pistol-grip " pattern.

These types are all similar in their essential principles, and the blade fixtures are also very similar to the type which has been previously fully described. There may be a slight saving in weight with the tubular type as compared with the older pattern.

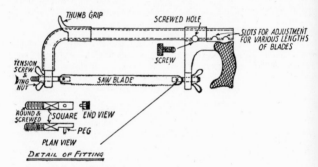

Fig. 44.—Tubular-frame Hack-saw (modern type).

Hack-saw for Sheet-metal Work (Fig. 45)

This hack-saw differs from those which have been previously described. In the cutting of sheet metal the depth of the saw-cut is often deeper than that for other classes of work. Further, it is difficult with a blade of narrow width to cut to a straight line for a great depth.

This type of frame, therefore, comprises a sheet-steel body with " beading " around its outer edges, except at that edge immediately adjacent to the blade.

This sheet-steel plate is of no thicker construction than that of the blade itself. Therefore, it follows in the path of the slot which the blade cuts out, and it also forms a guide for keeping the blade cutting in a downward straight line. At the tension-screw end, the width of the sheet steel is far greater than that of the handle end. This is because most of the sawing is effected at the front end of the blade. Further, at the same position the distance from the blade to the stiffener frame exceeds by far that of an ordinary type hack-saw frame, thus allowing further penetration of the blade during its sawing process. The steel plate keeps the whole framework rigid, so in certain cases the hack-saw can be used held in the right hand only, in a manner somewhat similar to a butcher's meat-saw. This allows the user free use of his left hand for holding or steadying the work-piece while it is being sawn.

This type of hack-saw caters for only one blade length, there being no provision for various lengths of blade. For the larger blade sizes, the width of the sheet-steel plate at the front end is considerably increased, and that of the

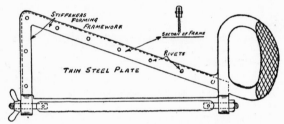

Fig. 45.—Hack-saw for Sheet-metal Work.

handle end may be increased in proportion. The type illustrated in Fig. 45 is only a small one, and is fitted with a fine blade suitable for the cutting of sheet brass.

HAND-SHEARS (Fig. 46)

These are similar to a strong pair of scissors, and are sometimes called "tin-snips", but, unlike household scissors, they have short "jaws" or "blades". Hand-shears are also of much stouter construction than scissors, and have long handles which turn inwards at their ends.

These tools are used for cutting thin sheet metals, etc. Those which are used for cutting "straight" sheets have straight jaws, but those which are used for the cutting of cylindrical or curved sheetwork have curved jaws.

Hand-shears are usually held in the right hand, whilst the

left hand grips the work-piece in order to steady it. They can, however, be used for the cutting of fairly stout sheet metal, by firmly securing one of the handles in a vice with the shear-jaw projecting slightly beyond the vice-jaw end, the unsecured handle of the shears being used as a lever for operating the movable shear-jaw.

These tools are very handy and useful for cutting small pieces of thin sheet metal, brass, copper, aluminium, zinc, etc.

If, on attempting to cut sheets, these are found to be too thick or too strong, a hack-saw should be used. If the sheet is too wide or unsuitable for the application of a hack-saw, possibly a cold chisel and a hand hammer could be used.

Where hand-shears are used, or even where an attempt to use them is made, on metals which are too hard or too thick, their jaws are liable to be damaged. Such instances may

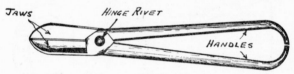

Fig. 46.—Hand-shears.

easily arise where a beginner is not acquainted with the duty which hand-shears are designed to perform. With injudicious use, the joint rivet of the hand-shears may also be strained.

After prolonged use the cutting edges of the jaws become blunted and they can be resharpened by grinding, preferably on a wet grindstone. The cutting edges should be ground and returned to an angle of approximately eighty-five or eighty-six degrees.

After regrinding the blade jaw-faces, one should ascertain that these faces meet properly. If they do not meet but have a space or gap between them when in the normal closed position, the blades will not cut properly.

After extensive or prolonged use the " hinge-pin " or rivet, by which the two blades are secured together, is sometimes found to be slack. This slackness will cause inefficient cutting. The rivet can easily be tightened by holding one end of it on a solid steel block and hammering the other end. It is important to see that there is no " side play " or side movement at the hinge-rivet position, yet it must not be hammered up too much, or it may be made too tight to facilitate easy operation.

Hand-shears are supplied in overall sizes which range from six inches up to approximately fifteen inches in length.

Similar tools, but with their jaws arranged to cut in a parallel motion, are made. These are often called " bolt-croppers ", and are chiefly used for cutting off small bolt heads which are " rusted up " and therefore difficult to unscrew.

The bolt-cropper is occasionally used for cutting stout sheet metal, but it is more frequently used for cutting sheets of " expanded " metal or " mesh-type " steel reinforcement for concrete work. The bolt-cropper should not therefore be confused with hand-shears or, as the latter are often called, " tinman's snips ".

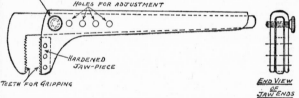

Fig. 47.—Adjustable Pipe-wrench.

ADJUSTABLE PIPE-WRENCH (Fig. 47)

This is the tool which is used for holding small- to medium-sized round objects, such as gas or water-pipes, or solid round bars.

Gas pliers for the holding of small, round objects were described earlier in this chapter, but any round bars or pipes which are too large in their diameter to be effectively gripped by gas pliers, should be held by an adjustable pipe-wrench. From the illustration, it will be seen that there are a number of holes in which to place the peg or set-screw after opening out the jaws by sliding them along; the jaw-faces can thus be arranged with varying spaces between them.

Imposed in them the jaw-faces have saw-edged teeth which effect a sure grip on any object that is being held.

These " teeth " often slightly damage the surface of the

object which is gripped, especially if great force is applied, but the " damage marks " can easily be erased by gently filing them down when the job is completed.

This particularly applies to pipework, when a pipe-wrench has been used to screw-up pipe fittings, such as its unions, sockets, etc.

These tools are very effective in use and handy for dealing quickly with any medium-sized object of a round shape that it is desired to rotate or screw-up. For pipework of a larger size, however, other types of tools would be used, and as these are of a special character they are dealt with in a later chapter.

ENGINEER'S VICE (Figs. 48 and 48a)

A vice is a tool with which an object is held in position whilst work is performed upon it. It is usually secured to a stout work-bench, and positioned at a suitable height to enable the work to be done in a standing position. An engineer's vice differs from a carpenter's chiefly because it is made entirely of steel or iron and steel. It is also of stouter and heavier construction than a carpenter's vice.

A vice is composed of two jaws, one of which is fixed to the framework, and the other adjustable. The adjustment is effected by the moving jaw being mounted on a sliding-screw arrangement. The screw is operated by the hand-turning of a suitable handle.

One of the earliest types of vice to be used, and one which is still used, is the " leg-type ". This tool, as will be seen from the illustration, has one long leg which extends down-wards towards the floor. The fixed jaw projects upwards from the top of the leg. Near the base of the leg is a hinged fixture, from which the arm projects and on which is mounted the movable jaw.

A screwed spindle passes through this arm at a position just below the jaw formation. By turning this screwed spindle, the jaws are opened or closed. A flat, cranked-type leaf-spring which opens the jaws when the spindle is unscrewed is positioned between the leg and the movable jaw arm.

The leg-type vice is sometimes called a " blacksmith's " vice, and is frequently seen in the village blacksmith's shop. It is of stout construction, and in most cases the jaws are made of solid steel. Some older types, however, have jaws which are made of wrought iron ; and in some cases the whole vice, including the leg and the arm, may be made of either wrought iron or forged steel.

The leg-type vice has one particular disadvantage— the moving jaw, which is pivoted at the lower end of the spring position, moves through an arc, and, therefore, when the jaws are in a comparatively " wide-open " position, their faces are not parallel to each other. Consequently, when any object which necessitates their being opened a considerable amount is secured between the jaws it is gripped by a wedge-like action.

When work of such a nature as to cause vibration, such as hammer-and-chisel work is performed on an object held

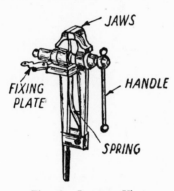

Fig. 48.—Leg-type Vice.

in this way, the object is inclined to work loose owing to the wedge action or taper of the jaw faces. In some cases the object or work-piece may completely " jump out " of the vice-jaws.

Of recent years, therefore, much attention has been devoted to the manufacture of what are termed " parallel sliding " jaw vices. A typical vice is illustrated in Fig. 48a. This type is designed so that whatever position the sliding-jaw may be, whether it is opened a small or large amount, its face is dead parallel with that of the fixed jaw.

It will be appreciated that this principle has a great advantage over that of the leg-type vice. Furthermore, the modern type of vice also has its screw mechanism entirely enclosed. This prevents any particles of grit or filings from getting lodged between the spindle screw-threads, and therefore lengthens their life.

In addition to the foregoing refinements and improvements, some modern vices have " quick-release " mechanisms fitted to their screw arrangements. By means of these quick releases the jaw grip can be released instantaneously by merely pressing a release catch which is situated adjacent to the handle end.

This mechanism releases the " screw-working " mechanism and allows the moving jaw to be slid in or out, irrespective of turning the screw handle, but the latter can still be used in the ordinary manner when required.

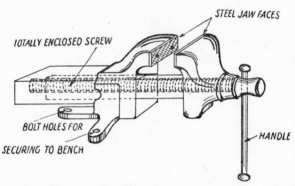

Fig. 48a.—Parallel Sliding-jaw-type Vice.

There are many types of parallel sliding-jaw vices now sold, some of which incorporate the quick-release method of operation.

Most modern vices have renewable jaw-faces. These are steel plates which are secured to the jaws by countersunk-headed set-screws. The plates are usually " checkered " or have small recesses of a " check " pattern impressed on their work faces. These are designed to grip the object in the vice jaws. If, however, undue pressure is brought to bear on the object these " impressions " are transferred to it.

In order to safeguard an object which is of a semi-finished nature from being damaged when it is thus secured in a vice, lead clamps are frequently used. These are merely sheets of lead of a suitable size and thickness which are placed between the vice jaws, turned down over the tops of the jaws, and are thus shaped to somewhat resemble the latter. The lead, which is of a softer nature than the steel of the jaw face-plates, does not damage or mark any finished or semi-finished work-piece.

As an alternative to lead, copper, tin, or zinc may be used for forming clamps, but these are not quite so effective as lead (see Fig. 48b). Also Fig. 87b for application.

Clamps are frequently used in conjunction with a vice for holding in position brass or gunmetal linings of bearings, during the chipping of oil-grooves or the drilling of grease-holes, etc. This is done after the liners have been turned or machined in a lathe, and it is therefore important not to damage the finish.

Clamps are also often used when any finished turned bolt requires its threaded length extending or its end has to be drilled in order to fit a split-pin, etc.

Fig. 48b.—Lead Clamps.

All vices should be securely bolted or screwed down to a strong work-bench. The bolts should be frequently inspected in order to ascertain that they have not worked loose, for the wood of which the bench is made often " gives " and allows the bolts to " bite " into it.

The size of a vice is denoted by the width of the jaw-face. Quite common sizes range between three inches and eight or even nine inches. The size of the vice naturally depends upon the class of work for which it is required.

Some special kinds of vices are now made with " swivelling " heads. The block comprising the jaws is mounted on a swivel arrangement, thus enabling it to rotate in a horizontal plane through a semicircle or even a complete circle, yet it can be securely " locked " in any intermediate position when so desired. This type of vice provides for the holding of an awkwardly shaped work-piece which could not be gripped in an ordinary vice because a projecting part of it may foul the bench or some other part of the vice.

A vice requires very little attention, beyond keeping it

reasonably clean and giving its screw mechanism an occasional drop of oil.

When the jaw-face plates become badly worn they can be replaced by unscrewing the set-screw which holds them in position.

PNEUMATIC-TYPE VICE (Fig. 48c)

This type of vice is operated entirely by compressed air. It comprises a base plate to which is attached a fixed jaw at the position indicated in the illustration.

Mounted on the base plate is a compressed-air cylinder with a plunger, which also forms the other jaw. At the position shown in the illustration there is an air-supply connection, which is coupled to an air-supply pipe. The latter has a foot-operated-type valve, which is used for

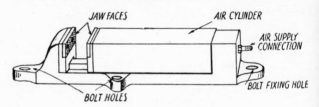

Fig. 48c.—Pneumatic-type Vice.

regulating the supply of air to the cylinder. This action thus controls the jaw spacings. An outlet valve for releasing the pressure on the jaws is also foot-operated. By this system both hands are left free for use by the operator, which is a distinct advantage.

The vice is suitable for work-pieces up to approximately two and a half or three inches in width. It is also of a portable nature, and for light classes of work it need not be bolted to a bench, but may be rested on a stand or work-table.

PUNCHES, DRIFTS, ETC.

Centre-punch (Fig. 49)

The " centre-punch " is one of the commonest forms of punch used in everyday engineering work. It is made of cast steel which has been tempered, and its point is ground to a fine, conical shape.

A centre-punch serves many useful purposes. One of its chief uses is for making a small " pin-point " mark in the exact position where a hole is to be drilled in any work-piece.

This procedure assists the drill at the commencement of its cutting or drilling operation. In addition to giving a

HAND GRIP

SHARP POINT

Fig. 49.—Centre-punch.

" lead " to the drill, it also prevents the drill point from " straying " away from the exact position during its starting process.

In use a centre-punch is held firmly between the thumb and first two fingers of the left hand in a vertical position. The point is carefully placed in the exact spot where the

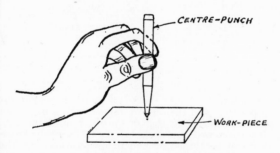

CENTRE-PUNCH

WORK-PIECE

Fig. 49a.—Hand-grip for Centre-punch.

mark is to be made, and the " striking " end is given a light blow from a hand hammer held in the right hand (see Fig. 49a).

For the drilling of holes of small diameter, it is usual to make one centre-punch mark dead in the centre of the desired drill-hole position, but for holes of larger diameter it is a more common practice, after forming the central point mark, to " scribe " a circle of equal diameter to the drill

C

and to put a centre-punch mark at four cardinal points as shown in the illustration (Fig. 49b).

This procedure forms a better guide for the " machinist " or " driller ", and enables him to see that the drill circumference and the drill centre keep to the desired positions as the drill penetrates the work-piece.

In addition a centre-punch may also be applied in cases where a small " set-screw " has broken off at a place " flush " with the face of the work-piece in which it has been screwed. The procedure in such cases is to form a small " indentation " by use of the centre-punch at a position near the outer edge of the broken screw. Next, the

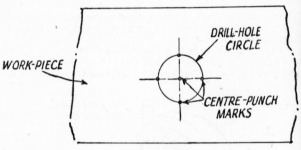

Fig. 49b.—Centre-punch Marking for a Large Drill-hole.

punch point is fixed into the small punch hole, but in a sloping position, and then gently tapped with a hammer in an opposite direction to that of the rotation of the screw when it was screwed into position. In this way the broken set-screw can be turned around in its position, and thus unscrewed sufficiently to raise its end just above the surface of the work-piece so that it can be gripped by pliers and completely extracted (see Fig. 49c).

A further use for a centre-punch is that of " marking " objects. In engineering work cases occur where it is essential to have certain components, or parts of machinery which are intended to be assembled in their correct order or sequence, marked in some way.

Let us consider, for example, two half-steps or " half-brasses " of a bearing. These may be, to all intents and purposes, identical with each other, but one of them may

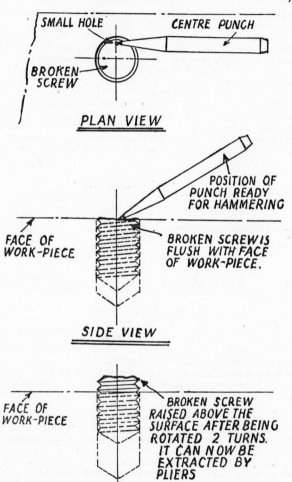

SMALL HOLE CENTRE PUNCH

BROKEN SCREW

PLAN VIEW

POSITION OF PUNCH READY FOR HAMMERING

FACE OF WORK-PIECE

BROKEN SCREW IS FLUSH WITH FACE OF WORK-PIECE.

SIDE VIEW

FACE OF WORK-PIECE

BROKEN SCREW RAISED ABOVE THE SURFACE AFTER BEING ROTATED 2 TURNS. IT CAN NOW BE EXTRACTED BY PLIERS

Fig. 49c.—Extracting Broken-off Screw by Aid of Centre-punch.

have been fitted or bedded down to the bearing base, and the other fitted to the bearing cap.

During the dismantling of the bearing both the half-brasses would be marked so as to ensure that each, on its being replaced, will be fitted into its respective part. A centre-punch mark is therefore made on the ends of each half-brass, with the brasses' relative " faces " contacting each other.

Should it, however, happen that several similar bearings of any one machine are to be dismantled, one would be marked as described above, whilst the second bearing would be marked with two centre-punch marks, each adjacent to the other, at one end of each bearing brass or half-brass

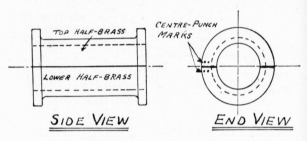

Fig. 49*d*.—Marking " Bearing-brasses " by Use of a Centre-punch.

(see Fig. 49*d*). The third bearing would be given three such marks on each half-brass, and so on.

This process is often adopted when motor-engine pistons have to be dismantled, in order to ensure that each piston, on its being reassembled, is fitted correctly into its relative cylinder.

A modern type of centre-punch is the " automatic ". This type has an internal spring mechanism and as the top part of the punch comprises a plunger no hammer is required.

The punch is placed in position on the work-piece, and as its point contacts the latter, pressure is applied to the top by the hand. This closes up the internal spring until it reaches an automatic " catch ", and the spring is released. The lower part, which contains the punch-point, is encased in a sliding sleeve, and when the spring catch is thus released a small " blow " is transmitted to the point, thus effecting a mark on the work. As every blow is uniform, so is each resulting mark which the point makes.

This type of punch can often be obtained with variable adjustment, and by altering the position of the release-catch on the spring the force of the blow can be regulated or varied.

The fact that this automatic punch can be used to deliver blows which give marks of uniform size is very useful in cases where a long line may have been scribed on work. In order to define this line clearly a series of punch marks of uniform size can be made all along it.

This method is very useful in cases where the scribed line may become defaced or even obliterated during the transport of the work from one workshop to another.

Pin-punch (Fig. 50)

A " pin-punch " somewhat resembles a centre-punch,

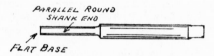

Fig. 50.—Pin-punch.

except that it has no sharpened point, nor does its lower end taper. Instead of tapering the latter is parallel, and it has a round, flattened end. The pin-punch is chiefly used for driving out the pins or rivets of small roller chains, in the absence of a proper extractor tool. It is also used for driving out " dowel-pegs " which are of a parallel pattern. If used for chain-pin extraction work, the lower end of a pin-punch should be hardened in order to stand up to its duty efficiently. It should also have its lower end of sufficient length to penetrate the required depth of the work for which it is employed, i.e., if it is used for extracting a chain link-pin, it must be at least as long as the width of the chain on which it is to be used, for very often such chain link-pins or rivets cannot be released unless they are driven completely through the links.

For that class of work it is often advisable to use a punch of a slightly smaller diameter than that of the chain link-pin. Further, in chain construction, these link pins, after being fitted, are slightly burred or riveted over at each end. It is therefore advisable to either grind or file down one end of the pin until it is " flush " with the outer face of the side link. This will greatly facilitate the pin's removal. It is also far easier than " drifting " it in its previous state, because the burred head would then have to be distorted until it was small enough to be driven through the holes of both the side links and the roller.

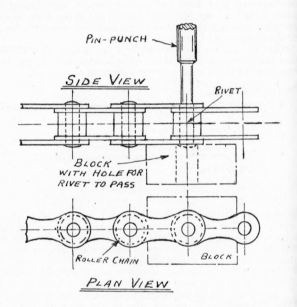

Fig. 50a.—" Drifting Out " a Chain Rivet by Pin-punch.

It is essential to have the chain link resting on some steel object which has a hole in its centre of slightly larger diameter than that of the rivet. This is to allow for the chain link to be well supported, whilst space is provided for the rivet to be effectively driven out (Fig. 50a).

A nut can frequently be used for this operation, provided, of course, that it is of ample depth.

Drift-punch (Fig. 51)

This is similar to a pin-punch, but its lower shank tapers. It is generally of heavier and stouter construction than a pin-punch, although drift-punches vary considerably in size according to the nature of the work for which they are used.

A drift-punch is used for " drifting " out tapered pins from work-pieces. In addition, it is frequently used for starting to extract a rivet from a chain, then once the rivet has commenced to move the drift-punch is exchanged for a pin-punch, which, by reason of its having a parallel shank, can be used for driving the rivet right through the chain link. A drift-punch is frequently used for sinking nail heads below the surface of timber-work after the nails have been driven home by hammering flush with the timber

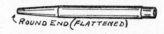

Fig. 51.—Drift-punch.

surface. Furthermore, in cases where a bolt is in position through two or more parts of metal (or timber and metal), it may be found, when the nut is removed, that the bolt shank is " binding " in its position. A drift-punch may then be applied to drive the bolt through the holes. This sometimes occurs in machinery which has been in use over long periods, and which may have become rusted.

Drift-punches vary in design, and for some classes of work they may have cranked or curved forms. They may be of either cylindrical or rectangular cross-section.

Key-drift (Fig. 52)

This tool is possibly one of the commonest forms of special drift-punch. It is used for driving out keys from gearing which is keyed to shafting. Instead of having a straight shank, it is slightly cranked in order to allow its lower end to lie flat in the key-way from which it is desired to extract the key. Its lower part is made of rectangular cross-section of suitable size for the key on which it has to be applied. The top part may be of either round or rectangular cross-section.

A key-drift is used, in cases where the key-head position

is difficult of access, to extract the key from the head end by inserting a " wedge " between the head and the boss of the wheel. A key-drift can be inserted at the narrow taper end of the key, and by striking the former with a hammer the key can be extracted (Fig. 52a).

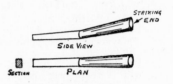

Fig. 52.—Key-drift.

Should the gear boss, however, be in such a position that it is close to a bearing or another gear-wheel, etc.—which thus leaves very little space for driving out the key completely, after loosening it—the gear boss may be slid on its shaft towards the key-drift position, which then gives full access to the key for its easy removal from the shaft key-way

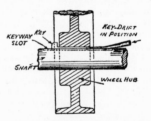

Fig. 52a.—Application of Key-drift.

CALIPERS

These are tools or instruments which are used for measuring objects, but with only a fair degree of accuracy. There are two chief types of elementary calipers, and both are used for measuring the diameters of articles. They are made of a mild grade of steel, and are composed of two legs which are lightly riveted (between two washers) together at one end so as to form a pivot or hinge. The method employed for measuring with calipers is to " span " the free ends of the legs either over or inside the object whose diameter it is desired to measure.

When the tips of the caliper legs just comfortably touch the sides of the object the caliper is removed and placed alongside the divisions of a steel measuring rule. The dimension is subsequently " read off " from the rule. This method is a quick and ready way of ascertaining the dimension of any object with a fair amount of accuracy.

Calipers vary in size according to the size of work for which they are required.

Note.—For obtaining the precise or accurate dimensions of objects in engineering, measuring instruments which are termed " *precision* " *instruments* are used. These are chiefly " vernier " calipers and " micrometers ". Such instruments are mainly concerned with machine-shop practice, which is dealt with in Volume III.

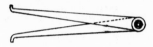

Fig. 53.—Inside Calipers.

Inside Calipers (Fig. 53)

" Inside calipers " consist of two straight legs, each of which has its extreme end bent outwards. The legs at their wide ends are fixed or secured together by a rivet which is sufficiently slack to allow movement of the legs when they are gently tapped or knocked on any hard object, but of such a tightness as to prevent the legs from moving apart freely.

Inside calipers are used for measuring the internal diameters of tubular or hollow-shaped objects. In use the tips of the legs are spaced to the approximate dimension of the tube " bore " diameter by gently tapping the legs apart, then a trial is made by inserting the calipers inside the object (see Fig. 53a).

In carrying out this operation great care must be taken to ensure that the caliper leg-tips are inserted evenly. Furthermore, care must be taken to see that the leg-tips are placed at the widest part of the tube's diameter, or an error in the measurement will result. If, on inserting the legs as just described, their tips do not quite touch the inner wall faces of the tube, the calipers must be extracted and their tips must be opened out slightly.

The operation of inserting into the tube must be repeated

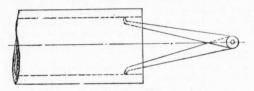

Fig. 53a.—Use for Inside Calipers.

and adjustments made to the leg-tips until the latter just comfortably touch the inner faces at the widest part of the tube. This having been successfully accomplished, care must then be taken in extracting the caliper legs, for if one

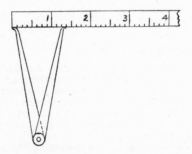

Fig. 53b.—" Reading Off " the Dimension from a Steel Rule.

leg is slightly knocked the resulting dimensions will not represent the true diameter of the tube.

Provided the calipers have been successfully withdrawn, their leg-tips are next laid along a steel rule and the measurement can be read-off (Fig. 53b).

A beginner should be very careful when withdrawing the caliper legs, for they are apt to " bind " on the wall faces inside the tube in positions which are not the tube's widest part.

Further, during the withdrawal of the instrument great care should be taken to ensure that one leg-tip does not advance in front of the other, or either leg-*bar* might be inadvertently knocked against the tube end, thus upsetting the dimension reading for the true diameter. Before finally withdrawing the leg-tips it should be ascertained that they are not touching the tube walls too tightly. If they are strained, through a certain amount of " springiness " in their legs, immediately on their release from the tube they will return to their normal position, and an incorrect reading will result.

After a little practice the beginner will soon acquire the correct " touch " for both inserting and withdrawing the calipers.

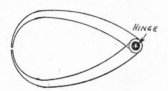

Fig. 54.—Outside Calipers.

Outside Calipers (Fig. 54)

These are used for measuring the outside dimensions of either rectangular or circular-shaped articles, and if used with care will give results of a fair degree of accuracy.

Outside calipers have legs which are curved throughout their entire length, as shown by the illustration. The legs are fixed together at their wide ends by a rivet, in a similar manner to that which has already been described for inside calipers. The leg-tips at their extreme ends are flat.

To use outside calipers the tips are spaced to the approximate dimension of the object and they are then applied to it. If they span the object too loosely, the legs are gently tapped a little closer together and again applied. This operation is repeated until the caliper leg-tips just comfortably touch both sides of the object without straining or springing the legs (Fig. 54*a*).

Should the article which is being measured be of a

cylindrical nature, care must be taken to ensure the legs are placed at dead right angles to the object, and not crossways, or a wrong dimension will result. It is usual, when it is believed the desired touch has been obtained, to move the caliper-legs once or twice lightly in and out of position over the object to ensure that they do not grip it too tightly or too loosely. In doing so, care must be taken to retain the leg-tips as square as possible with the face or centre line of the object. Once the user is satisfied that he has successfully spanned the object, the caliper-legs are finally removed, laid against the divisions of a rule, and the dimension read off.

Should the caliper-legs become unduly loose or slack after

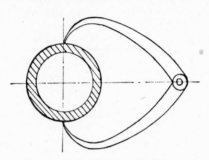

Fig. 54a.—Use for Outside Calipers.

prolonged use, they can easily be tightened by either nipping the hinge rivet in a vice or gently tapping the rivet head with a hammer.

ENGINEERS' COMPASSES OR DIVIDERS (Fig. 55)

These instruments are made of steel. Those illustrated are made of " spring " steel. Compass points are usually hardened in order that they may be ground to a sharp point. This enables them to stand up to the duty which they are called upon to perform over long periods.

There are various types of compasses used in engineering work, but all have some means of regulating the opening of the legs so as to allow for fine adjustments. In the type illustrated the " spring-head " forms the opening movements, and it is regulated by the " wing nut " which is mounted on the small, screwed spindle. By turning the wing nut the points of the legs can be brought closer together, and if the wing nut is unscrewed the spring forces the legs apart.

Engineer's compasses are used for scribing circles and curves, during the marking out of machinery or components, in order to show the machinist the places to be machined. Engineer's compasses are therefore used in a similar manner to the artist's or draughtman's lead or pencil compasses. They are also used as " dividers " for spacing out or dividing lines on a work-piece.

Owing to the hard nature of materials used in engineering work, the compass points scratch a mark on the surface. In order that these scratch marks may be more easily seen, it is usual to either rub the surface of the work-piece with

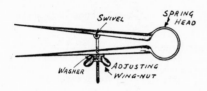

Fig. 55.—Engineer's Compass or Dividers.

white chalk, or to coat with a white-wash brush. This is most essential with surfaces of such metals as cast iron, steel, etc., which are of a dark colour, and unless " whitened " the scratch marks cannot easily be seen. It is of less importance to whiten materials such as aluminium, as it marks more easily, and scratches can readily be seen, especially if the aluminium is in sheet form.

To use the compasses for scribing a circle or a curve, it is usual to make a small, fine, centre-punch hole in which to place one leg of the compass. The other leg is used for scribing the arc or circle. The compass should be held by one hand in a similar manner to that employed when using an ordinary pencil compass, but more pressure has to be used on the scribing leg in order to form the scratch-mark and make it penetrate well into the surface of the work. If one particular compass leg is constantly used for scribing work it will, in due course, get worn down. Later, when the legs are closed up for scribing arcs or circles of very small radii, it will be found that they are of unequal length, and difficult to use. It is therefore a good plan to use each leg-point alternately, so that both wear down an equal amount.

In due course, when the compass points require sharpening, they should be ground, preferably on a " wet " grindstone.

JENNY OR "ODD-LEGS" (Fig. 56)

This tool is really a special kind of compass. It consists of one straight leg, which has a sharp point at its lower end,

Fig. 56.—" Odd-legs " Calipers or Compass.

and one leg which is similar to that of an inside caliper. The legs of a " Jenny " are secured together by washers and a rivet as in the case of calipers.

" Odd-legs " can be used for several different purposes,

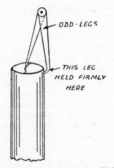

Fig. 56a.—Shaft Centre Obtained by Using " Odd-legs."

but perhaps one of their chief uses is for finding the centre of any circular-shaped object such as a shaft. In order to find such a centre, the leg having the curved tip is reversed so that the curve points inwards towards the pointed leg.

The former is then held at the outer edge of the shaft's perimeter and is used as a pivot, while the pointed leg is used for scribing an arc as shown in Fig. 56b (see Figs. 56a and 56b). The arc can be "struck" or scribed anywhere near the estimated centre of the shaft, but it is advisable to make it of a fair length, as shown in the illustration.

Next, carefully keeping the odd-legs spaced the same distance apart, the instrument is moved to a point exactly opposite to that from which the arc was scribed. From this position, a second arc is made which intersects the first.

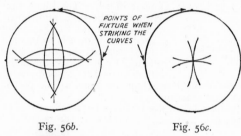

Fig. 56b.   Fig. 56c.

Arcs Defined by " Odd-legs."

At a point on the shaft's circumference ninety degrees from this point, a third arc is struck, and a fourth arc is made from a striking point exactly opposite to that of the third position.

We have now four arcs scribed out, of which each pair intersect each other. By joining these points of intersection with two straight lines, it will be seen that the point where the latter cross, or intersect, is the true centre of the circular shaft.

An alternative method for determining the shaft's centre is illustrated in Fig. 56c. It is not easy to obtain exact results, however, until after having had some little experience.

The beginner may not judge accurately the distance for striking the first arc. This should not cause anxiety, as when the subsequent arcs are struck (providing each is made with the instrument legs spaced the same distance apart) there will be a small, somewhat rectangular figure between the four arcs. The corners of this figure can be joined by two diagonals, whose point of intersection will be the centre of the shaft (see Fig. 56d).

A further use for odd-legs is for scribing a line inside a tubular-shaped object. In order to effect this the legs of

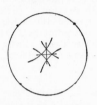

Fig. 56d. — Alternative Method of Finding Centre of Shaft.

the instrument must, first of all, be reversed, for the pointed leg in this case is used as the pivot, or centre, from which the line is scribed. Furthermore, in order to do this some means of support must be provided. This can be accomplished by inserting a wooden block into the end of the tubular object. The block should be driven into the tube, and tightly fixed at a depth equal to that where the desired scribe-mark is to be made.

The centre of the object must next be found, by any of the methods which were described for a solid shaft, and a

mark must then be made in the wood at the point which has thus been determined (see Fig. 56e).

The pointed leg of the odd-legs is placed at this position and a point is marked in the tube at the necessary depth

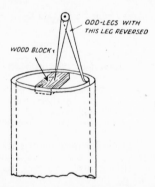

Fig. 56e.—Scribing a Line Inside Tube Surface by "Odd-legs."

from its end from where the mark is to be scribed. In order to do this, the outer leg is pressed against the inner wall of the tube, and whilst holding the odd-legs by the hinged end, the arc or circle is struck. Care must be taken to keep the hand in the same vertical plane exactly over the centre mark while the arc is being struck, in order that the scribed mark is made parallel to the end of the tube. This can later be checked by measuring down from the tube's end at one or two positions well apart from each other. When the arc has been successfully made, the wooden block can be " prised " out of the tube from the same end it was inserted, or it can be driven out from the opposite end.

In addition odd-legs can be used to scribe on a rectangular object lines which are parallel to its side edge. In order to effect this, the bent " leg-tip " is positioned along the edge of the object. The pointed leg is used as a scriber, and the

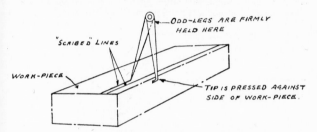

Fig. 56f.—" Odd-legs " Used for Scribing a Line Parallel to Side Edge of Rectangular Object.

odd-legs are moved bodily along, care being taken to keep both legs parallel and the curved leg in close contact with the edge of the object (see Fig. 56f).

CHAPTER II

GAUGES : LIMIT GAUGES, WIRE GAUGES, SCREW-THREAD GAUGES

A " GAUGE ", in engineering work, is a special kind of measuring instrument or tool. There are several different forms, each of which has some special use or application.

During the manufacture of an object it frequently has to

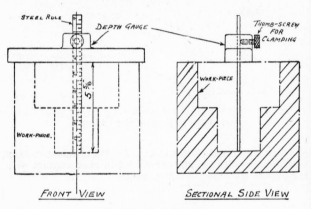

Fig. 57.—Depth Gauge.

be " tried " or compared with a pattern in order to make it to a pre-determined size or shape. Gauges are used for this purpose. They are therefore carefully and accurately made, so that they may form a " standard " to be worked to.

DEPTH GAUGE (Fig. 57)

For objects which have flat-based recesses or internal shoulders, a simple method of ascertaining the depth is to insert a steel rule and measure the distance from the end of the object to the recess. This may be effected with a fair amount of accuracy.

In order to obtain a greater degree of accuracy, a " depth gauge " should be used. This instrument consists of a " boss " with two arms projecting from it, one at each side. The arms are of sufficient length to " span " the object concerned. Through the centre of the boss projects a narrow, but rigid, steel rule. This is held in position by means of a " thumb-screw ", and the steel rule can be moved through the boss any desired amount by sliding it. It can also be " clamped " in any desired position, by screwing up the thumb-screw.

The depth gauge is placed over the end of the article, the thumb-screw is slackened off, and the steel rule slid vertically down the recess until its end rests on the face or shoulder of the recess. While the rule is in this position it is clamped or " locked " by tightening up the thumb-screw. The instrument is then withdrawn, and the depth of penetration read off from the rule at the lower face of the boss.

Some types of depth gauge, instead of having a steel rule, are fitted with a round steel bar. In this case the amount of penetration is ascertained by measuring off the length of the steel bar against that of a steel rule.

PLUG GAUGE (Fig. 58)

This is made of solid steel, and is of cylindrical form. At one end is a handle. It is used as a " template " or " pattern " to which any hollow, cylindrical-shaped object may be made. The object is machined to the size of the gauge, and the latter is occasionally " tried " to see if it can be inserted into the object during the machining process. Plug gauges are therefore made to given nominal sizes, and are each used for their own particular size of bore diameter.

It will be readily appreciated that a solid-steel shaft which has been accurately machined to a nominal finished dimension of, say, two inches diameter, cannot be readily fitted to the hub or boss of a wheel of exactly the same diameter. Therefore, either the shaft must be made very slightly smaller than the nominal two inches or the wheel hub must be bored slightly larger than the nominal two inches diameter. In order to do this with uniformity, an agreed " limit ", which has been adopted by certain engineering institutions as " standard ", must be worked to.

In the case of the shaft and wheel hub, let it be assumed that the limit to be worked to in both cases is one thousandth part of one inch. The shaft would therefore be made to a finished size of two inches, minus one thousandth of an inch. The wheel hub would be bored correspondingly larger, by one thousandth of an inch, over and above the nominal two inches diameter bore.

Therefore the " plug " gauge used by the machinist for boring the wheel hub would, in all probability, be two inches plus one thousandth of an inch. It would be termed 2 inches + 1 " thou ", the latter being the adopted abbreviation for " one thousandth of an inch ". The shaft would be made correspondingly less, and would thus be indicated on the drawings as 2 inches — 1 " thou " diameter, or 2 inches — 0·001 inch. In fixing the hub to the shaft by a key, the latter would be driven tight, thus leaving no

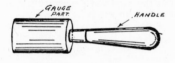

Fig. 58.—Plug Gauge.

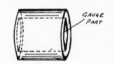

Fig. 59.—Ring Gauge.

" clearance " once the whole unit was " keyed up ". The reader will now fully realise the importance of using plug gauges for such work. The use of standard gauges thus caters for instances where two different engineering firms make two separate parts. For example, an electrical engineering firm may supply a motor with a shaft of a nominal two inches diameter. Another firm may make the belt-pulley for the motor drive. Unless the same agreed " limit " is worked to by both firms, difficulty may be experienced during the assembly of the two parts. A plug gauge is therefore made to a very accurately finished size. It is sometimes called a " male " gauge.

RING GAUGE (Fig. 59)

This gauge, as its name infers, is of " ring " or hollow cylindrical shape. It is used for just the reverse operation to that of the plug gauge. The ring gauge, therefore,

would be used in order to try out the shaft in the case which
has been previously mentioned.

It is frequently termed a " female " gauge, and is often
used in conjunction with a male gauge of the same specified
" nominal " size. Both gauges should be used with great
care to ensure that the working surfaces do not get
" bruised ". When not in use they should be coated with
oil or grease to prevent corrosion.

GAP GAUGE (OR HORSESHOE TYPE) (Fig. 60)

This gauge is, in some respects, a combination of the plug
and ring gauge. It consists of a gauge for " external "

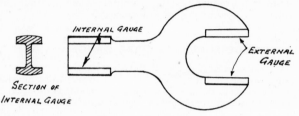

Fig. 60.—Gap Gauge.

work at one end, whilst the opposite end is used for
" internal " work.

The " gap " gauge is used in cases which do not lend
themselves to the application of a ring gauge. For instance,
the gap can readily be fitted or " tried over " a shaft while
it is still in a lathe, whereas to use the ring gauge, the end
of the shaft must be free in order to slip the gauge over it.
The gap gauge is not quite so accurate in use as the
cylindrical type, but it serves a very useful purpose.

It will be appreciated that the use of all the foregoing
gauges in a machine shop is very essential for repetition or
mass-production work which involves the manufacture of
many similar articles or components. The use of gauges in
the making of components ensures rapid assembly of the
various parts when the complete machine is assembled.
They thus save considerable time in the fitting up of
machinery, which was previously done by manual labour in

the fitting shop by filing up each individual part to fit its corresponding part.

The systematic use of gauges makes for simplicity if spares are required for any machine after it has been in prolonged use. If the spare parts are manufactured to the same gauges as those used for the original parts, little difficulty is experienced in fitting up the replacements.

Formerly it was the practice to leave the machining of components, more or less, to the discretion of the machinist, who worked to a nominal dimension. The job of obtaining a perfect fit was then left for the fitter. Modern practice is therefore to cut out, as much as possible, all hand-fitting work by adopting the general use of fine limits and gauges in the machine shop. Of recent years this has been the principle adopted by all the leading automobile manufacturers for their mass-production systems, and this principle is used extensively in America.

GAUGE PLATE (OR WIRE GAUGE) (Fig. 61)

A " gauge plate " is a hard, flat steel plate which has slots formed all around its edges. The slots are cut out and are ground to accurate specified sizes. Each slot is numbered, and the number represents a certain dimension which has been agreed as a Standard by the British Standards Institution.

A gauge plate is used to determine the thickness of certain steel, brass, or copper sheets, etc., or the diameter of various wires, etc. Usually the thickness of steel plates of three-sixteenths of an inch and upwards is simply given in inches or fractions of an inch.

For plates less than three-sixteenths of an inch in thickness, it is usual to refer to them as a given number, although steel plates of one-eighth of an inch in thickness, and sometimes those of one-sixteenth of an inch, are often referred to in fractions of an inch instead of the equivalent number. However, in practically all cases the thickness of very thin plates is referred to by giving a number.

The chief use of the gauge plate is to ascertain the thickness of a steel sheet or the diameter of a wire in the finished state. The very small sizes of sheets and wires, etc., so closely resemble each other in their thicknesses, that it is often impossible to distinguish them unless they are tried in the slots of a gauge plate. When using the latter, care must be taken to keep the slot sides or walls dead parallel with those of the sheet or wire which is being measured.

The gauge-plate slot should be gently slipped over the article it is desired to measure. Force should not be used to push the slot over the article, nor should the latter be too loose in the slot, or a wrong result will be obtained.

There are several types of gauge plate in use; each one representing some particular gauge. Among some of the most common gauges used in engineering are: the Imperial Standard, the Birmingham Wire Gauge, the Birmingham Sheet Gauge, Stubs Wire Gauge, S.W.G. (Standard Wire Gauge), Warrington Wire Gauge, Whitworth Wire Gauge, etc.

The number denoted by each type of gauge differs slightly in size, so when using a gauge plate great care must

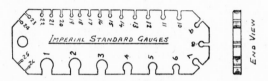

Fig. 61.—Gauge Plate or Wire Gauge.

be taken to ensure that it is the correct one specified for a particular job. The S.W.G. is chiefly used for brass and copper. For typical tables, see Fig. 61*a*.

The British Standards Institution is the authority in this country which is responsible for the fixing of the various "standards" to which firms engaged in the engineering industry agree to work. They thus form a co-ordination between the manufacturers and users of the products.

By the agreed adoption of such standards, the efficient manufacture of interchangeable parts (or components) is effected.

In order to avoid misunderstandings as to which particular gauge was intended to be used for specific jobs the British Standards Institution in 1914 adopted the following sizes for the thicknesses of thin iron and steel sheets.

FIG. 61a.

Iron or Steel Plates (Birmingham Gauge).

Descriptive number	Equivalent thickness in inches.	Descriptive number.	Equivalent thickness in inches.
15/0 B.G.	1·0000	20 B.G.	0·0392
14/0 B.G.	0·9583	21 B.G.	0·0349
13/0 B.G.	0·9167	22 B.G.	0·03125
12/0 B.G.	0·8750	23 B.G.	0·02782
11/0 B.G.	0·8333	24 B.G.	0·02476
10/0 B.G.	0·7917	25 B.G.	0·02204
9/0 B.G.	0·7500	26 B.G.	0·01961
8/0 B.G.	0·7083	27 B.G.	0·01745
7/0 B.G.	0·6666	28 B.G.	0·015625
6/0 B.G.	0·6250	29 B.G.	0·0139
5/0 B.G.	0·5883	30 B.G.	0·0123
4/0 B.G.	0·5416	31 B.G.	0·0110
3/0 B.G.	0·5000	32 B.G.	0·0098
2/0 B.G.	0·4452	33 B.G.	0·0087
1/0 B.G.	0·3964	34 B.G.	0·0077
1 B.G.	0·3532	35 B.G.	0·0069
2 B.G.	0·3147	36 B.G.	0·0061
3 B.G.	0·2804	37 B.G.	0·0054
4 B.G.	0·2500	38 B.G.	0·0048
5 B.G.	0·2225	39 B.G.	0·0043
6 B.G.	0·1981	40 B.G.	0·00386
7 B.G.	0·1764	41 B.G.	0·00343
8 B.G.	0·1570	42 B.G.	0·00306
9 B.G.	0·1398	43 B.G.	0·00274
10 B.G.	0·1250	44 B.G.	0·00242
11 B.G.	0·1113	45 B.G.	0·00215
12 B.G.	0·0991	46 B.G.	0·00192
13 B.G.	0·0882	47 B.G.	0·00170
14 B.G.	0·0785	48 B.G.	0·00152
15 B.G.	0·0996	49 B.G.	0·00135
16 B.G.	0·0625	50 B.G.	0·00120
17 B.G.	0·0556	51 B.G.	0·00107
18 B.G.	0·0495	52 B.G.	0·00095
19 B.G.	0·0440	—	—

Extract from B.S. No. 350—1944, Conversion Factors and Tables are given by permission of the British Standards Institution, 2 Park Street, London, W.1, from whom official copies can be obtained, price 5s. post free.

FEELER GAUGE (Fig. 62)

This gauge resembles a common " pocket knife " in shape and size. Unlike a pocket knife, which usually has two " blades ", a feeler gauge has several. These blades are all secured to one common rivet-type hinge, as shown in the illustration. The thickness of each blade is equal to a certain number of thousandth parts of one inch.

These, in engineering, are commonly called " thous ". The number of " thous " which each blade represents is usually inscribed on that particular blade. This gauge is used as a measure between any two flat surfaces. It is

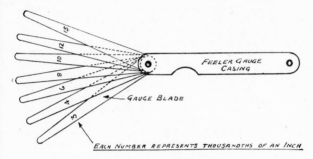

Fig. 62.—Feeler Gauge.

frequently used for setting valve-tappet clearances in internal-combustion-engine work. These allow for the expansion of the valve stems when the engine becomes hot.

In general engineering a feeler gauge is also used for ascertaining the amount of wear which may have taken place between any two moving parts of machinery.

To use this type of gauge, the required blade is opened out from the casing in a similar manner to that used for opening a pocket knife. The other blades are left unopened in the casing, and the open blade, bearing the number of " thous " which it is desired to use, is gently pushed between the surfaces of the parts concerned. Should this particular blade be found to be loose or slack, it is folded back into its casing, and the next larger size is tried.

This operation is repeated until a blade just fits comfortably between the two surfaces without any undue slackness or tightness. When this condition is arrived at it is obvious that the distance between the two surfaces is that which is represented by the number on the blade. Thus, if the two surfaces which are being examined represent those parts of a machine which have been operating for some considerable time, the number on the blade of the gauge represents the number of thousandths of an inch of wear which has taken place—providing of course the parts were a touching fit to each other originally.

To use a feeler gauge for setting the valve clearances of a motor-car engine, however, it is usual to pre-determine the amount of clearance to be given to each valve tappet. The exhaust valves, which are subjected to the hot gases from the explosions in the cylinders, usually require larger working clearances than those of the inlet valves. Some types of engines require three thous clearance for their inlet valves, whilst their exhaust valves may require six thous clearance. In these instances the feeler blades marked " three " and " six " would be used. While the engine is " cold " (*i.e.*, before it is started up and allowed to get warm or hot) the tappets are adjusted by using these gauge blades as the standard or template to which the valve tappets are set. When a comfortable working fit for the gauge blade has been obtained between each valve and its corresponding tappet, the locking nut of the latter is tightened up so as to retain the tappet in position for the desired clearance.

After completion of the tightening-up process, it is usual to insert the feeler blade again so as to ensure that in tightening it, the tappet did not move or turn slightly, as the least movement will often upset the delicate adjustment to which the tappet has been set.

SCREW-THREAD GAUGE (Fig. 63)

This tool resembles the feeler gauge in its general shape and size. A screw-thread gauge, however, has blades of equal thickness, and the blades also have teeth spaced along one edge, as is shown in the illustration. The blades are made of tough steel, and the teeth are accurately formed to the size and shape of the screw-thread which each represents.

Different types of screw-threads vary in both their pitch and formation, yet some types, especially those of the fine thread, closely resemble each other on casual examination. Because it is often difficult to distinguish one type of thread from another, a screw-thread gauge has to be used. If a gauge-blade of a known screw-thread is fitted into the spaces between the teeth of the thread in question, the type of the latter can be ascertained.

The screw-thread gauge must be inserted with the length

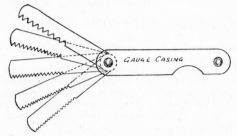

Fig. 63.—Screw-thread Gauge.

of its blade parallel to the length of the threaded article which is under test (see Fig. 63*a*). This type of thread gauge, in addition to being used for ascertaining the kind of external threads of an object, can also be used for testing

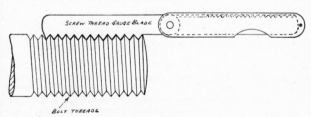

Fig. 63*a*.—Application of Screw-thread Gauge.

internal screw-threads, such as those of pipe connections, etc.

Furthermore, a screw-thread gauge can be used for checking the depth of threads when a " chaser " is being used for thread cutting in conjunction with a lathe. When used in this way the gauge forms a useful guide for deciding to what extent pressure should be applied to the chaser.

CHAPTER III

THREAD-SCREWING TOOLS

THESE are used for forming screw-threads on articles by various hand methods. They comprise, chiefly, two forms: those which are used for internal thread cutting and those used for external thread formation. Modern methods of thread cutting for mass-production work mostly involve the use of machine tools, but hand methods are employed, especially in cases which call for only a small quantity of work to be done.

SCREW TAP (Fig. 64)

" Screw tap " is the name of the tool which is used for cutting screw-threads internally in various metals. Naturally, a tap must be made of metal which can be hardened to a very high degree in order to enable it to cut threads in other metals which are comparatively " softer " and which have also been previously drilled with holes of a suitable size for the tap to enter.

" Screw taps " are supplied in sets of three. The one which is used for the first operation is called the " Taper " tap, and the tap which is next used in order of sequence is called the " Second " tap. The third tap, and the one which finally completes the cutting of the screw-thread formation, is called the " Plug " tap. Each of the three taps, when used in its proper sequence, performs a certain amount of work towards the effective cutting of the thread formation.

A screw-tap is of cylindrical formation, but usually has a square head formation. The " shank " or main body of a tap has screw-threads formed upon it. They are in three segments, and the spaces or " flutes " between these segments are for the purpose of allowing the small pieces of cut metal to be collected from the work-piece during the cutting of the threads. The small pieces of metal thus cut out are called " cuttings ". The square end at the head of a tap is used to attach the handle or " tap wrench " as it is called.

The taper tap has fully defined screw-threads only at the

top end of the shank. The formation of these threads or teeth gradually diminishes towards the lower end of the shank, as shown in the illustration.

The second tap greatly resembles the taper, but it has a longer length of more fully defined teeth along its shank; and has only a comparatively short length of taper.

The plug tap has fully defined cutting teeth formed on practically the whole length of its shank.

To use these taps, the taper tap is pushed into the hole in which it is desired to form the threads. The hole must be of a slightly smaller diameter than the tap. The tap is then lightly struck at the top end to force the taper well

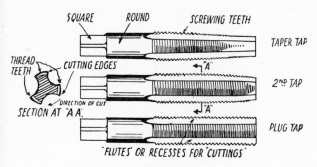

Fig. 64.—Screw Taps.

into the hole, also to make the teeth formation bite into the sides of the hole. Next, the tap wrench (Figs. 64a and 64b) or handle is fitted to the square head of the tap. If the hole which is to be threaded is required to have right-handed threads, naturally right-handed taps must be used. In this case the tap wrench used would be turned in a clock-wise or right-handed direction in order to commence work.

As the wrench is gripped by both hands for the turning operation, it should also have pressure applied to it in a downward direction in order to press the tap farther into the hole when the screw-threads of the tap begin their cutting action. It is helpful when starting to tap the hole (i.e., the term used when screw-threads are formed) to add a little oil to the tap teeth, as this greatly assists their cutting action.

The rotation of the tap wrench is continued, and at each turn more downward pressure should be applied. As the work proceeds and the tap penetrates into the hole deeper, it should occasionally be withdrawn and the used oil, together with the bits of " thread cuttings ", wiped off with a clean rag. Fresh oil should then be applied; the tap replaced in the hole, and the " tapping " proceeded with.

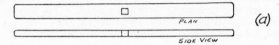

Fig. 64a.—Tap-wrench (small type).

When the threads of the taper tap have passed completely through the hole in the object, the tap should be extracted.

Next the second tap is applied to the work, and the screwing operations which have just been described are repeated. The second tap has more clearly defined cutting threads, which cut out the threads more definitely than those of the taper tap.

After the second tap has been passed through the hole

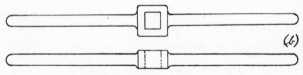

Fig. 64b.—Tap-wrench (large type).

successfully, it is withdrawn and replaced by the plug tap, which has still more clearly defined threads. After its use the threads will be completed and the " set-screw ", stud, or bolt can be fitted.

If the object which is to be screw-threaded has only a recess or " sinking ", instead of having a hole completely through it, great care must be taken to ascertain when the tap has penetrated to the base of the sinking. This can be done, first, by measuring the depth of the sinking, then marking the measurement on the shank of the tap, either by a chalk mark or red-lead mark. This procedure gives a guide as to when the sinking base is being approached, and thus prevents the user from continuing to try to turn the tap when its base is resting on that of the sinking. If such a precaution is not taken, the tap will probably be broken, for a tap, especially if of only a small diameter, is very brittle. When tapping threads in recesses or sinkings a beginner is strongly advised to make full use of the measuring and marking process which has just been described.

Once some little experience has been gained, however, it will be found that the user can tell by the " feel " of the wrench when the tap has reached the base of a sinking. The " touch " or " feel " is not so easy to distinguish when using taps of small sizes, so, in such cases, it is always advisable to make use of the marking process. It is also essential to clean the taps of cuttings, and the recess should be cleaned out more often when sinkings are being screw-threaded.

The whole of the foregoing descriptions, advice, and instructions refer solely to articles which are to be screw-threaded while they are secured in a vice and are, therefore, positioned so that the hole or sinking is in a vertical plane.

Frequently, however, tapping has to be done in a horizontal or inclined direction, especially to castings of a large size. In such instances the taps should be " entered ", pressed in, and pressure applied during the rotation of the wrench in exactly the same ways as those mentioned earlier. The cleaning and oiling of the taps should also be similarly carried out.

The chief point of tapping holes in a vertical plane, and also those of an inclined nature, is to be sure that the pressure is exerted on the tap wrench in such a manner that it ensures a direct, straight thrust into the hole. Pressure must not be exerted on one side of the hole only; this is often the tendency when holes are tapped in positions other than those on a vertical plane.

The " hand grip " of a tap wrench should somewhat resemble that shown for draw-filing in Fig. 37, but after a semicircle has been completed by the wrench, both hands should be released and replaced on the opposite handle ends.

Having dealt with the formation of screw-threads in hollow objects by hand methods, the formation of external threads by manual operations may now be considered.

STOCK AND DIE (Fig. 65)

" Stock and die " is the tool which is used for the formation of external screw-threads on objects such as round bars of metal and pipes, etc. In addition, a stock and die is often used for extending the thread formation of bolts lower down their shanks.

A stock consists essentially of a framework which is recessed in order to accommodate the die. The latter is either in one piece or consists of two halves. The type which is shown in Fig. 65 consists of the die of two half formations. The stock or framework has a handle at each end of the centre-piece, and "guides" on which the die-

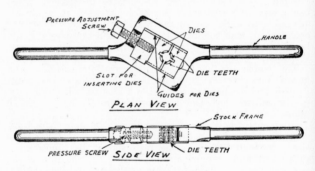

Fig. 65.—Stock and Die.

pieces can slide along when they are placed in position. A stock is usually made in such a manner that several sizes of dies can be accommodated in it.

Pressure on the die is obtained by tightening the adjustment screw, which presses the half-die (to which it is adjacent) closer to the other, thus effecting more cutting pressure on their teeth.

For the formation of screw-threads, the object is first of all secured in a vice in a suitable position, with the end which is to be threaded projecting a few inches above the vice-jaws. It is sometimes found useful to file the projecting end slightly, to give it a taper, which thus gives a lead on which to fix the dies (see Fig. 65a).

The stock and die is then applied, the halves of the latter being so spaced that the teeth just span the work-piece. Sufficient pressure is applied to the adjusting screw to cause the die-teeth to grip the work. Next, a little oil should be applied to the die-teeth.

For screwing right-handed threads, the stock handles are turned or rotated in a clock-wise direction. During this rotating process downward pressure should be applied in the same manner as when tapping.

After a few revolutions of the stock have been made a little more oil may be required. The turning of the stock is then continued until the thread is formed down the object for the desired length.

The stock is then rotated in the opposite direction until it reaches the top of the work-piece or its starting point. At this stage it can be removed and cleaned of the small pieces of thread cuttings, and the work-piece should also be cleaned of cuttings and used oil.

The stock is then replaced on the work and the adjusting screw tightened up a little more. The operation is repeated by rotating the stock down the work-piece. This will have the

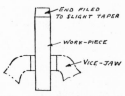

Fig. 65a.—End of Work " Taper-filed " to Give a lead for Die.

effect of cutting the threads deeper and is repeated until the screw-threads are clearly defined or completed.

This can be checked by a trial of the nut for which the object is being screw-threaded. Should the nut be found to be too tight, the work-piece should be given further applications of the stock and die, with a little more pressure added to the adjusting screw, until the nut will travel along the work-piece freely.

During this final screw-threading operation, however, care must be taken to avoid using too much pressure on the adjusting screw, which may result in cutting off too much metal, and thus causing too loose a fit for the nut.

It is very essential when using stocks and dies to clean away the metal cuttings frequently, otherwise they tend to wedge in between the teeth of the dies and damage the threads on the work-piece.

It is also essential to apply pressure to the adjusting-screw very gradually. If too much pressure is applied at once, there is a tendency to " tear off " parts of the threads from the work instead of gradually cutting away a thin layer, or " peeling " at each operation. Similar damage

D

will often result if the stock is rotated too quickly (see Fig. 65b).

These last two remarks particularly apply to the screw-threading of mild steel or brass. The best results are obtained when a nice, even pressure is applied to both

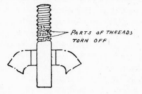

Fig. 65b.—Result of Applying too much Pressure to Die.

handles of the stock and the latter is rotated at a steady pace.

The hand grip for the type of stock illustrated in Fig. 65 should be similar to that which has previously been referred to for tapping work. The stock and die illustrated is the type which is used for work of approximately one-quarter of an inch up to about three-quarters of an inch in diameter.

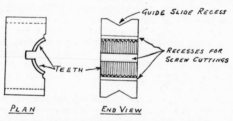

Fig. 65c.—Enlarged View of Half Die.

Larger sizes of stocks and dies of a similar type are also used for work of one inch upwards to two inches or so in diameter.

Enlarged details of a typical die (or half die) are shown in Fig. 65c, from which it will be seen that recesses are provided in which the metal cuttings collect.

The die teeth are formed for cutting in whichever direction they are rotated. After the work-piece has been

threaded down its length, further pressure can be applied
to the teeth before rotating the stock in an anti-clockwise
direction and bringing it to the top of the work-piece.
Screw-threading thus proceeds when the dies are rotated in
either direction.

It will be realised, therefore, that after running the stock
and die down the work, it is not necessary to bring the tool
to the top of the work—or starting point—before increasing
the cutting pressure by tightening the screw. This is a
great advantage, and saves much valuable time.

It is inadvisable to increase the cutting pressure midway
along the work-piece, for if this is done during the downward
run of the tool, and then, on reaching the " bottom " position of the work, more pressure is added, it will be found that
during the upward run the dies may tend to tear off the
threads on reaching the midway position. It is obvious

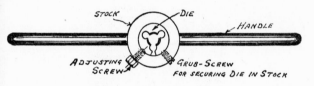

Fig. 66.—Stock and Die (for small work), Split-type Die.

that the cause of this is that the dies are called upon to cut
away double the usual amount of metal.

For the cutting of left-handed screw-threads, *i.e.*, those
which advance in an anticlockwise direction, special dies
are, of course, required. These, however, if of similar size
and shape, will fit into the same stock as is used for cutting
right-handed threads. The only real difference when using
left-handed dies is that the stock in which they are held is
rotated in an anticlockwise direction during the downward-
run or " stroke ".

The foregoing remarks apply to stocks and dies for work of
average normal sizes. Other types of slightly different
designs are used for small- or large-diameter work-pieces.
These types will now be dealt with in succession.

Stock with Single " Split Die " (Fig. 66)

This is used for the threading of round bars of a small
diameter, say, those between one-eighth of an inch, and one-
quarter of an inch. The die is usually of a shallow-

cylindrical form, and it has teeth formed in its centre, as shown in the enlarged diagram of Fig. 66a. At three points, adjacent to the screw-cutting teeth, clearance slots are formed in which the cuttings are collected.

This type of die has only one single adjustment slot. Pressure from the adjusting screw reduces the width of the slot and thus regulates the pressure on the teeth.

The die is sunk into the recess of the stock, and is held in position by means of a small " grub-screw ". It will be readily appreciated that the " die " can easily be changed for one of another size in a matter of seconds by simply unscrewing the grub-screw and releasing the pressure-adjusting screw.

This type of stock and split-die is used extensively for fine thread-cutting work of the " close-pitch " type. It is

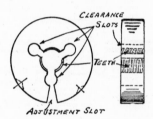

Fig. 66a.—Split-type Die (enlarged view).

a very handy tool for the formation of threads on cycle- or motor-cycle-wheel spokes or articles of a similar size. Dies for this type of stock can be obtained for either left-handed or right-handed screw-threading formation.

Screw Plate (Fig. 67)

This tool is a special form of fixed dies. It is used for very light or small work, and comprises a hard steel plate with a handle at one end only. The plate has several perforations, each of which is screw-threaded, and each perforation has two clearance slots, which are positioned one at each side as shown in the illustration. These perforations are usually in sets of three for each individual screw-thread size. Number one is used for the first screw-cutting operation, which partially defines the screw-threads on the work-piece. Number two perforation is then applied, and run down the work-piece to increase the cut. It is then withdrawn and replaced by number three. The latter, on its being run down the work, finally completes the screw-thread formation.

Should it be found, however, that after application of number three die perforation the work-piece is still a little tight for its corresponding nut, the third die can again be applied. This time, however, during the " running-down " operation the plate can be tilted *very* slightly out of the horizontal plane. This has the effect of causing the die-teeth to bite into the work a shade more, and will often give the desired result. Care must be taken to ensure that the inclination of the plate is not too great, or the screw-threads

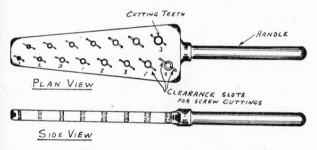

Fig. 67.—Screw Plate (or Die-plate) used for very light work.

will be damaged, or may even be ripped off completely. As no means of adjustment are provided for closing in the teeth of a screw-plate the tilting method is the only one which can be used as an inducement to the teeth to bite deeper into the work-piece. The screw-plate must, of course, be cleaned frequently to clear away the cuttings.

Large-type Stock and Die (Fig. 68)

For the formation of screw-threads on large diameters, heavier and larger types of stocks and dies are used. These are frequently supplied with four handles, as shown in the illustration. This type is intended for use by two or more operators, and is used extensively for the screw-threading of large pipework ranging between three and six inches in diameter.

The tool consists of a cylindrical-shaped, hollow centre-piece in which the dies are positioned. The latter are usually formed in three or four segments and are arranged so as to slide inwards or outwards from the centre. This sliding movement is controlled by a lever, which thus regulates the cutting pressure of the teeth. This lever is

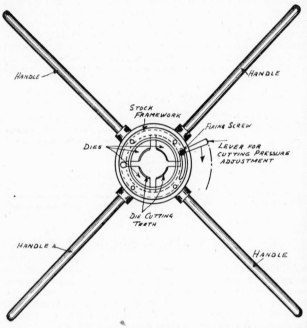

Fig. 68.—Stock and Die (large type).

secured in any desired position by a fixing-screw or locking-screw.

The four handles are secured to the stock centre-piece in bosses or lugs which project from the latter. Handles for this type of stock are often as much as three feet in length, and are made of solid, round steel bars. The dies are retained in position and their sliding movement is controlled by dowel pins.

Fixed over the dies is a removable cover plate which is secured in its position by two or more set-screws. When it is desired to change the dies for those of another size, the cover plate is removed from the stock centre-piece, and the existing dies removed by lifting them out of their recesses. This type of stock is usually made to accommodate several different sizes of die diameters.

It is usual to secure the work-piece in a horizontal or inclined position in a pipe vice. This enables the operators to stand each side of the handle arms and pull them around in turn. (For heavy-duty work two operators may pull on a handle together in one combined operation.)

The work-piece having been secured in the desired position in a vice (or special pipe-vice—if it is a pipe which is to be screw-threaded), the dies are then opened out so as to span the work. The whole tool is then applied to the work-piece, and the die segments are closed up until they engage the work-piece. Pressure is next brought to bear on the die teeth by lightly knocking the adjusting lever. As the teeth grip into the work the die segments are locked in that position by screwing up the locking-screw.

The handles of the stock can now be rotated and the screw-thread cutting started. When the thread screwing has proceeded for the desired length down the pipe, the locking-screw can be released, and the adjusting lever can be " knocked " again to close up the die segments slightly more.

After the locking-screw has been tightened the stock handles can be rotated in the opposite direction, and screw-cutting proceeds during this upward or backward direction of rotation.

These operations are repeated in their correct sequence until the threads have been successfully formed on the work-piece. During these operations oil should be added periodically to assist the screw-cutting action of the dies, and the latter should be cleaned occasionally of the cuttings, which readily collect in the recesses.

It is sometimes necessary with this type of tool, in order to thoroughly clean all the movable parts, to dismantle periodically the whole mechanism, as these cuttings have a habit of penetrating into the slots of the slides, as well as into the die-teeth recesses.

Hand Chaser (Figs. 69, 70 and 71)

Although this is actually a hand tool, it is used for work in conjunction with a lathe, which is a machine tool. A hand chaser is used for cutting screw-threads on objects or for more clearly defining threads which have been partially cut out or defined. There are two types of chasers : the external type is used for cutting external threads on objects, and the internal one is used for screw-thread cutting inside hollow objects.

A chaser is made of hard steel, and teeth are formed at its cutting end. At the opposite end of a chaser blade, a wooden handle is fitted, which resembles a file handle. The teeth of a chaser are accurately made to the standard sizes specified for each particular screw-thread. A work-piece to be screw-threaded is fixed securely in the lathe in a horizontal position and is then rotated.

If the chaser is pressed hard against the object and moved at a uniform speed in a horizontal direction along the face of the work, the pattern of its teeth will be cut or defined on the object. Furthermore, if the chaser is moved to and

CUTTING TEETH

Fig. 69.—Internal Hand Chaser.

fro along the object several times, the chaser's teeth cutting deeper each time, the full " profile " or " outline " of these teeth will in due course be completely reproduced on the object. This process is one of the chief uses of a chaser and, incidentally, it is also the elementary principle of a screw-cutting lathe, which is dealt with in Volume III.

For work-pieces which have only partly defined screw-threads on them a chaser can be used in the prescribed manner for effectively completing the thread formations, but in this case only very light pressure should be exerted on the work. In order to ascertain when sufficient material has been cut off, the work can be tested either by application of the nut to which it is to be fitted or by checking with a screw-thread gauge.

After prolonged use a chaser's teeth may be re-formed by accurately regrinding them, using the corresponding screw gauge as the pattern. If necessary, the chaser blade end can be softened, and after carefully filing the re-formed teeth to their correct shape, it can be re-hardened ready for use again. For screw-thread cutting of internal work, the chaser which is shown in Fig. 69 would be used, but the principle of operation is the same.

A chaser is a very handy tool where the existing screw-thread is too tight.

Typical external chasers are shown in Figs. 70 and 71. For the type shown in Fig. 71 it is claimed that the cutting

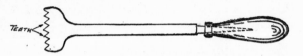

Fig. 70.—External Hand Chaser.

end is of more rigid construction and therefore less likely to yield or " give " when undue pressure is exerted on it.

A chaser is often favoured for thread cutting on brass-work. It is also frequently used for the cutting of very fine and small threads. This is possibly because any operation which is manually performed can be far more sensitive to the touch than that of a machine tool, which is more regular,

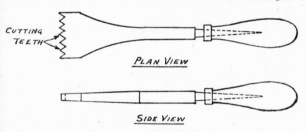

Fig. 71.—External Hand Chaser.

but less sensitive in its application to work of a delicate nature.

The art of using a chaser efficiently requires considerable skill and experience. These can be acquired only by constant practice, in order to get accustomed to the best cutting speed, and the most suitable pressure to be applied to each class of work.

Figs. 71a and 71b give tables of standard screw-thread dimensions. Fig. 71c gives a table of British Standard pipe threads which are used for general-engineering purposes.

Fig. 71a.

TABLE I. *British Standard Whitworth Screw-threads.*

Basic Sizes. B.S. Whit.

1	2	3	4	5	6	7	8
Nominal diameter, in.	Number of threads per inch.	Pitch, in.	Depth of thread, in.	Major diameter, in.	Effective diameter, in.	Minor diameter, in.	Cross-sectional area at bottom of thread, sq. in.
⅛ *	40	0·02500	0·0160	0·1250	0·1090	0·0930	0·0068
3/16	24	0·04167	0·0267	0·1875	0·1608	0·1341	0·0141
¼	20	0·05000	0·0320	0·2500	0·2180	0·1860	0·0272
5/16	18	0·05556	0·0356	0·3125	0·2769	0·2413	0·0457
⅜	16	0·06250	0·0400	0·3750	0·3350	0·2950	0·0683
7/16	14	0·07143	0·0457	0·4375	0·3918	0·3461	0·0941
½	12	0·08333	0·0534	0·5000	0·4466	0·3932	0·1214
9/16	12	0·08333	0·0534	0·5625	0·5091	0·4557	0·1631
⅝	11	0·09091	0·0582	0·6250	0·5668	0·5086	0·2032
11/16 †	11	0·09091	0·0582	0·6875	0·6293	0·5711	0·2562
¾	10	0·10000	0·0640	0·7500	0·6860	0·6220	0·3039
⅞	9	0·11111	0·0711	0·8750	0·8039	0·7328	0·4218
1	8	0·12500	0·0800	1·0000	0·9200	0·8400	0·5542

THREAD-SCREWING TOOLS 107

1⅛	7	0·14286	0·0915	1·1250	1·0335	0·9420	0·6969
1¼	7	0·14286	0·0915	1·2500	1·1585	1·0670	0·8942
1½	6	0·16667	0·1067	1·5000	1·3933	1·2866	1·300
1¾	5	0·2000	0·1281	1·7500	1·6219	1·4938	1·753
2	4·5	0·22222	0·1423	2·0000	1·8577	1·7154	2·311
2¼	4	0·25000	0·1601	2·2500	2·0899	1·9298	2·925
2½	4	0·25000	0·1601	2·5000	2·3399	2·1798	3·732
2¾	3·5	0·28571	0·1830	2·7500	2·5670	2·3840	4·464
3	3·5	0·28571	0·1830	3·0000	2·8170	2·6340	5·449
3¼	3·25	0·30769	0·1970	3·2500	3·0530	2·8560	6·406
3½	3·25	0·30769	0·1970	3·5000	3·3030	3·1060	7·577
3¾	3	0·33333	0·2134	3·7500	3·5366	3·3232	8·674
4	3	0·33333	0·2134	4·0000	3·7866	3·5732	10·03
4½	2·875	0·34783	0·2227	4·5000	4·2773	4·0546	12·91
5	2·75	0·36364	0·2328	5·0000	4·7672	4·5344	16·15
5½	2·625	0·38095	0·2439	5·5000	5·2561	5·0122	19·73
6	2·5	0·40000	0·2561	6·0000	5·7439	5·4878	23·65

* Dimensionally the ⅛ inch × 40 t.p.i. thread belongs more proportionately to the B.S. Fine series, but it has for so long been associated with the Whitworth series that it is now included therein.

† To be dispensed with wherever possible.

Extracts from B.S. Number 84—1940, Table 1, British Standard Whitworth Screw Threads, are given by permission of the British Standards Institution, 2 Park Street, London, W.1, from whom official copies can be obtained, price 7s. 6d. post free.

FIG. 71b.

TABLE 8. *British Standard Fine Screw-threads.*

Basic Sizes. B.S. Fine.

1	2	3	4	5	6	7	8
Nominal diameter, in.	Number of threads per inch.	Pitch, in.	Depth of thread, in.	Major diameter, in.	Effective diameter, in.	Minor diameter, in.	Cross-sectional area at bottom of thread, sq. in.
3/16	32	0·03125	0·0200	0·1875	0·1675	0·1476	0·0171
7/32	28	0·03571	0·0229	0·2188	0·1959	0·1730	0·0235
1/4	26	0·3846	0·0246	0·2500	0·2254	0·2008	0·0317
9/32	26	0·03846	0·0246	0·2812	0·2566	0·2320	0·0423
5/16	22	0·04545	0·0291	0·3125	0·2834	0·2543	0·0508
3/8	20	0·05000	0·0320	0·3750	0·3430	0·3110	0·0760
7/16	18	0·05556	0·0336	0·4375	0·4019	0·3663	0·1054
1/2	16	0·06250	0·0400	0·5000	0·4600	0·4200	0·1385
9/16	16	0·06250	0·0400	0·5625	0·5225	0·4825	0·1828
5/8	14	0·07143	0·0457	0·6250	0·5793	0·5336	0·2236
11/16	14	0·07143	0·0457	0·6875	0·6418	0·5961	0·2791
3/4	12	0·08333	0·0534	0·7500	0·6966	0·6432	0·3249
13/16	12	0·08333	0·0534	0·8125	0·7591	0·7057	0·3911

7/8	11	0·09091	0·0582	0·8750	0·8168	0·7586	0·4520
1	10	0·10000	0·0640	1·0000	0·9360	0·8720	0·5972
1 1/8	9	0·11111	0·0711	1·1250	1·0539	0·9828	0·7586
1 1/4	9	0·11111	0·0711	1·1250	1·1789	1·1078	0·9639
1 3/8	8	0·12500	0·0800	1·3750	1·2950	1·2150	1·159
1 1/2	8	0·12500	0·0800	1·5000	1·4200	1·3400	1·410
1 5/8	8	0·12500	0·0800	1·6250	1·5450	1·4650	1·686
1 3/4	7	0·14286	0·0915	1·7500	1·6585	1·5670	1·928
2	7	0·14286	0·0915	2·0000	1·9085	1·8170	2·593
2 1/4	6	0·16667	0·1067	2·2500	2·1433	2·0366	3·258
2 1/2	6	0·16667	0·1067	2·5000	2·3933	2·2866	4·106
2 3/4	6	0·16667	0·1067	2·7500	2·6433	2·5366	5·054
3	5	0·20000	0·1281	3·0000	2·8719	2·7438	5·913
3 1/4	5	0·20000	0·1281	3·2500	3·1219	2·9938	7·039
3 1/2	4·5	0·22222	0·1423	3·5000	3·3577	3·2154	8·120
3 3/4	4·5	0·22222	0·1423	3·7500	3·6077	3·4654	9·432
4	4·5	0·22222	0·1423	4·0000	3·8577	3·7154	10·84
4 1/4	4	0·25000	0·1601	4·2500	4·0899	3·9298	12·13

Note.—It is recommended that for larger diameters in this series four threads per inch be used.

Extracts from B.S. Number 84—1940, Table 8, British Standard Fine Screw Threads, are given by permission of the British Standards Institution, 2 Park Street, London, W.1, from whom official copies can be obtained, price 7s. 6d. post free.

FIG. 71c.

TABLE 15. British Standard Pipe Threads (Parallel).
(For General Engineering Purposes). Basic Sizes. B.S. Pipe.

1	2	3	4	5	6	7	8
B.S.P. size	Number of threads per inch	Pitch, in.	Depth of thread, in.	Major diameter, in.	Effective diameter, in.	Minor diameter, in.	Cross-sectional area at bottom of thread, sq. in.
1/8	28	0·03571	0·0229	0·3830	0·3601	0·3372	0·0893
1/4	19	0·05263	0·0337	0·5180	0·4843	0·4506	0·1595
3/8	19	0·05263	0·0337	0·6560	0·6223	0·5886	0·2721
1/2	14	0·07143	0·0457	0·8250	0·7793	0·7336	0·4227
5/8	14	0·07143	0·0457	0·9020	0·8563	0·8106	0·5161
3/4	14	0·07143	0·0457	1·0410	0·9953	0·9496	0·7082
7/8	14	0·07143	0·0457	1·1890	1·1433	1·0976	0·9462
1	11	0·09091	0·0582	1·3090	1·2508	1·1926	1·117
1 1/4	11	0·09091	0·0582	1·6500	1·5918	1·5336	1·847
1 1/2	11	0·09091	0·0582	1·8820	1·8238	1·7656	2·448
1 3/4	11	0·09091	0·0582	2·1160	2·0578	1·9996	3·140
2	11	0·09091	0·0582	2·3470	2·2888	2·2306	3·908
2 1/4	11	0·09091	0·0582	2·5870	2·5288	2·4706	4·794
2 1/2	11	0·09091	0·0582	2·9600	2·9018	2·8436	6·351
2 3/4	11	0·09091	0·0582	3·2100	3·1518	3·0936	7·517
3	11	0·09091	0·0582	3·4600	3·4018	3·3436	8·780

Extracts from B.S. Number 84—1940, Table 15, British Standard Pipe Threads (Parallel), are given by permission of the British Standards Institution, 2 Park Street, London, W.1, from whom official copies can be obtained, price 7s. 6d. post free.

CHAPTER IV
DRILLS AND DRILLING

DRILLING is the term which is applied to the action of forming a hole in anything by means of a tool which penetrates the material with a combined forward and rotary motion.

The tool which performs this work does so by cutting away small pieces of the material as it advances through the work-piece. This tool is known as a " drill " or a " bit ", and the rotary cutting action proceeds spirally.

Thus the formation of a hole by " drilling " is distinct from that of making a hole in anything by a " punch ", which has the effect of simply pushing or forcing its way through an object by " brute force " alone.

The reader will therefore appreciate that there is a vast difference between the formation of a hole by drilling it and by punching. This difference plays a very important part in engineering work, for the action of drilling a hole does not damage the area of metal which is adjacent to the perimeter of the hole, whereas a punch, which forms the hole by its action of shearing or tearing its way through the work, damages a small area around the perimeter.

A drill neatly cuts the metal away in small flakes as it passes or progresses through the work, whereas a punch presses out or shears the metal in one solid piece, and thus leaves a hole in the material, representing the size of the punch which passed through it.

Incidentally, a punched hole has sides which taper slightly, whereas the sides or walls of a hole which has been drilled are truly parallel. Normally the cost of drilling holes is more expensive than that of punching them, but machine tools known as " multiple drilling " machines have recently been vastly improved, so that when a large quantity of similar-sized holes have to be formed, the cost of drilling them by this machine is only slightly more than that of punching them, yet a far better result is obtained and the material is undamaged.

There are in use, several types of drills and bits, each of which has its own specific application or purpose. Drills, on account of the special duty which they are called upon to perform, must be made of very hard material; they must also be of suitable shape and size to enable them to be secured in some type of socket or holder to which a rotary motion can be applied.

Different means are employed for holding the drill or bit in its socket. Some forms of drills have tapered shanks, and other types parallel shanks, with a slot at the extreme shank end through which a cotter can be fixed For hand work, the drills are secured in a " brace " or a " hand-drill ". Still another form, very similar to the hand-drill, is known as a " breast-drill ", and there are " electric " and " pneumatic " hand-drills.

Before considering details of the various sockets, drill holders, etc., let us first consider the drills or cutting tools.

Fig. 72.—Flat Drill.   Twist Drill.

TWIST DRILL (Fig. 72)

Perhaps one of the most common and up-to-date type is the " twist " drill. As its name implies, this type of drill takes the form of a " twist " or " helical spiral ". It has parallel sides, however, and " flutes ", or recesses of spiral shape, are cut out from its shank. These flutes form slots into which the drillings or cuttings can collect as the drill proceeds with its work.

The forward edge of each flute forms the cutting edge of the tool, as will be seen on reference to the illustrations. The drill is pointed at its cutting end to give a better start at the commencement of cutting or drilling.

At the opposite end the drill shank either takes the form of a taper or continues in parallel form. Some drill ends of the latter pattern are slotted to accommodate a cotter pin which retains this type of drill in position and which also prevents it from rotating in its holder.

The tapered type of drill shank end is usually made to

some standard form of taper. One of the commonest forms is that known as a " Morse " taper, so called after its originator or inventor. The Morse taper drill shank does not require a cotter because it is held in position by its own taper, and as more pressure is exerted, the tighter it is secured in position.

The adoption of an agreed standard enables the various makers of both drills and drill holders to manufacture these components so as to be interchangeable.

Normally, twist drills are made for right-hand drilling work, *i.e.*, the drill is designed with its cutting edges formed to cut when it is rotated in a clockwise direction.

Twist drills for average normal drilling work have their points ground to an angle of approximately sixty degrees. After prolonged use the drill point will require resharpening and this can be done with a special tool which secures the drill in such a position that the correct angle is maintained during its treatment on a grindstone or emery wheel, whichever is used. When resharpening twist drills the best results are probably obtained if a wet grindstone is used.

When drills are being used, some form of lubrication is essential. For hand-operated drilling machines, ordinary thin machine oil will be found quite satisfactory. This form of lubrication may also be found satisfactory for hand-operated small electric or pneumatic types of drilling machines, but for higher-speed drilling work, or in cases where heavier types of twist drills are used for high-powered drilling machines, it will be found essential to use a mixture of soft soap and water or soft soap, soda, and water. A copious supply of the mixture should be used, for it greatly assists the drill with its cutting work and also prolongs its life.

Care should be exercised not to apply too much pressure when using twist drills of only small diameter. This is very important, for if the pressure applied is sufficient to cause the drill to bend slightly it will very likely break. Should the break occur inside the work-piece, and below the surface, it is often a difficult matter to extract the broken part, which, being made of very hard steel, is naturally brittle, and in trying to extract it, more may be broken off lower down inside the work-piece. This only complicates matters.

When using small twist drills in any kind of hand-type drill holder every endeavour should be made to keep the drill in as straight a line as possible, for often quite a small divergence from the straight will cause the drill to break.

FLAT DRILL (Fig. 72)

The " flat " drill is one of the older forms. It usually has either a parallel shank or a tapered one with a cotter slot formed in it. This type of drill generally has a main shank which tapers gradually towards its pointed end. The latter then broadens out and is flattened.

Both sides of the flattened end are pointed, and subdivided by a small " notch " or slot as is depicted in the illustration. The angles of the points may vary between approximately thirty and forty degrees, depending upon the class of work for which the drill is to be used. Both the vertical sides and the inclined points are ground to form sharp cutting edges.

The flat drill, although slower in its cutting or drilling action than a twist drill, is quite efficient, but of recent years it has been largely superseded by the twist type.

When in use, the cuttings from the drilling operation

Fig. 73.—Countersink Drill.

escape by passing between the flats of the pointed end. The resharpening of its points and the sides of the flats requires more skill than the resharpening of a twist drill.

COUNTERSINK DRILL (Fig. 73)

" Countersink " is the term used when a bolt-head (or a screw-head) is recessed into the work-piece in such a manner that the head is " flush " with or does not project above the surface of the work. The term is usually applied to bolts which have inverted cone-shaped heads. In order to drill or form the specially shaped holes for accommodating such bolt-heads, a countersink type drill would be used and the process is known as " countersinking " the holes.

In order to effect this countersink, as is shown in the illustration, Fig. 74, two methods can be used. One is to use the point of a drill of approximately twice the diameter of that which was used to drill the plain or parallel-sided hole.

Another method which gives the same result is to use a

special countersinking drill. A typical formation of such a drill is depicted in Fig. 73. From this illustration it will be seen that the base of the drill is conical shaped and is provided with several tapered teeth which have cutting faces on one side only. The cutting faces are, of course, sharpened in the direction of the rotation of the drill.

The socket or holder end may be similar to any of those which have already been described, but for drills of modern type, the Morse-taper pattern is used most frequently.

For the efficient fitting of countersunk bolts or screws it is often preferable to form the countersink slightly deeper than that of the bolt or screw head to ensure that the latter does not project above the surface of the work-piece.

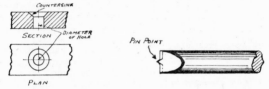

Fig. 74.—A Countersunk Hole. Fig. 75.—Flat-ended Drill.

FLAT-ENDED DRILL (Fig. 75)

Having dealt with the types of drills which are used for effecting a conical-shaped countersink, attention may now be paid to another type designed to accommodate screw heads (or bolt heads) of " cheese-headed " formation.

In engineering it may be essential to sink cylindrical heads of bolts or screws below the surface of the work-piece. A typical hole for such a bolt or screw will be seen on referring to Fig. 75a. A flat-ended drill, similar to that shown in Fig. 75, would be used in this instance.

It will be seen that the end of the drill is flattened, and that it has a small central point at its " tip ". From this point two flat cutting faces project; one at each side of the point. These cutting faces are sharpened in the direction of rotation.

In order to form a hole for a cheese-headed screw or bolt, make a centre-punch mark in the desired position, apply the drill point, and the drilling can be proceeded with to its required depth. On the completion of this procedure the drill is withdrawn. The " pin-point " from the flat drill's impression, which remains in the centre of the hole, is then used for the point of the twist drill, which is next used in order to drill the shank hole or main hole in which the bolt or screw shank is to be fitted. In addition to this type of cheese-headed hole being required for screws or bolts, it is extensively used for similarly shaped pins.

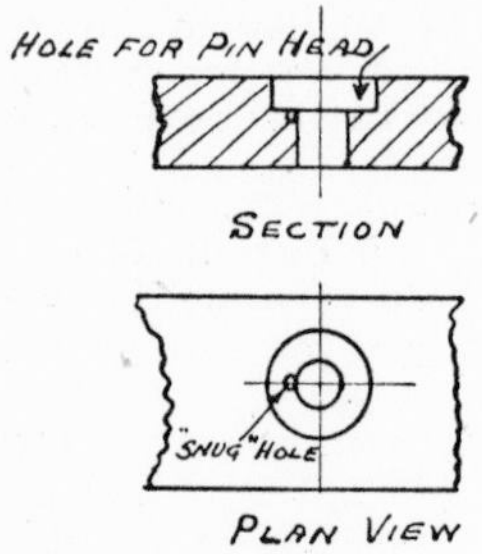

HOLE FOR PIN HEAD

SECTION

SNUG HOLE

PLAN VIEW

Fig. 75a.—Hole for Cheese-headed Screw or Pin.

Incidentally, at this stage it might be mentioned to advantage that whenever such screws are sunk below the surface of the work-piece, they are usually of the slotted-head type. The slot forms a means of securing the screw in position with the aid of a screwdriver. The same slotted-head formation is also used for bolts, as this provides a method of holding the bolt and preventing it from rotating during the process of tightening up its nut with a spanner.

For a cheese-headed pin, however, a " snug " is provided immediately beneath the head in order to prevent it from rotating in the hole. A snug is a small peg either driven or screwed into the pin. In order to accommodate the snug, a further small hole must be drilled in the pin hole at the position shown in Fig. 75a. More detailed particulars of this are mentioned in Volume II under the heading of components, etc.

CUTTING EDGES

PIN POINT

"MORSE" TAPER SHANK

Fig. 75b.—Type Used for Drilling Thin Sheet Metal.

As an alternative method a cheese-headed hole may be drilled by first using an ordinary twist drill to drill the bolt-shank hole, then a flat-ended drill of suitable diameter for the cheese-head. In this case, however, the pin forms the guide for the two holes thus being drilled concentric with each other.

A flat-ended drill of the type illustrated in Fig. 75b is often favoured for drilling holes in thin sheet copper or zinc, and other alloys.

SLOTTING DRILL (Fig. 76)

This type of drill is used for finishing off flat-based slots which have previously been formed by a flat-ended drill. As its name infers, the " slotting " drill is used for such work as " keyway " slots.

Fig. 76.—Slotting Drill.

In using a slotting drill, the drill-holder mechanism must travel in a horizontal direction, or if it is stationary or is fixed in a permanent position the work-piece must be moved in a corresponding horizontal direction. Whichever method

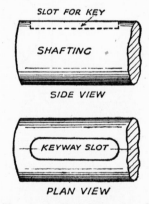

Fig. 76a.—Key-way in Shaft Formed by Slotting Drill.

is adopted, the travel must be of a uniform speed. Furthermore, it is essential that the " feed " or pressure applied to the drill should be kept uniform for the keyway's entire length. The feed or amount of cut should also be small, and only very gradually increased at each stage.

The chief difference between a slotting and a flat-ended drill (as will be noticed from the illustration in Fig.

76), is that the former has a recess at the centre of its base. This is because the original slot formation of the keyway has previously been formed by a flat-ended drill which has a point at its base for locating the slot centre. This is therefore unnecessary with the slot drill, as the location will have been correctly formed prior to its application.

A typical use of the slot drill is shown in Fig. 76a, where it is applied to the keyway of a shaft. A considerable amount of skill is required for this class of work.

SPECIAL CUTTING DRILL (Fig. 77)

In cases where holes of very large diameters are to be formed, a special appliance has to be used. This consists of a main spindle which has a slot through it near its lower

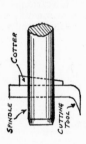

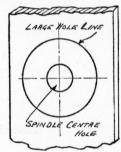

Fig. 77.—Special Cutting Drill (used for large holes).

Fig. 78.—Cutting a Large Circular Hole.

end. Into the slot is fitted a special cutting tool, as shown in Fig. 77.

This is secured in its position by a tapered cotter, and by means of the latter the radius of operation of the cutting tool can be fixed. The cutting tool can be adjusted and locked in position for a number of holes of various diameters.

This type of special cutting drill is used extensively for cutting holes of large diameter in steel plates, etc. The whole principle of operation is that the tool cuts out a disc of diameter equal to the required hole size.

The mode of application for this special cutting drill is as follows. First of all, a spindle hole must be drilled in the centre position of the desired large hole. The spindle hole must be of just sufficient clearance size to accommodate the drill spindle. Next, the cutting tool is adjusted so that the point of it is positioned at the perimeter of the large hole which is to be formed.

When the cutting tool has been positioned it is securely locked by means of the cotter which is tightly driven into the spindle slot and so wedges the cutting tool tightly in that position. The drill spindle is then lowered on to the work-piece until the cutting-tool point contacts the latter.

The spindle is then set in motion and the cutting tool begins to cut. As the latter rotates, more pressure is applied at each revolution, until the disc is completely cut out, thus leaving the large hole formed in the work-piece.

The tool, which is made of hard steel, can easily be resharpened on either a wet grindstone or an emery wheel.

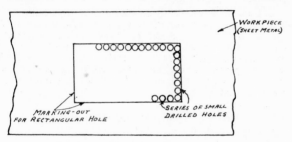

Fig. 79.—Forming a Large Rectangular Hole by Drilling a Series of Small Holes as Shown.

The application of this special cutting drill is illustrated in Fig. 78.

Incidentally, for the cutting of a rectangular-shaped hole in sheet metal, one method is to cut out the hole by using a hammer and cold chisel, which is fairly easy provided that the metal is only of a thin character.

Should the metal be too thick for this procedure to be adopted the following method can be used. First mark out the desired shape and size on the work-piece. A series of small-diameter holes can then be drilled all around inside and with their outer edges just touching the perimeter of the marked-out shape.

The edge of each of these small holes should nearly touch the edge of the adjacent hole. The metal which remains between can then be chipped away or filed through, and finally the rough edges can be filed down, thus leaving the hole formed truly to the desired shape (see Fig. 79).

The foregoing method is possibly somewhat laborious, but it often has to be resorted to. In a modern engineering works, however, the cutting of such a hole would no doubt be carried out by the employment of either an oxy-acetylene burner or one of the arc welding type, by which means a rectangular or irregular hole can be burned out in a very few minutes. This method will be more fully dealt with later on under the heading of welding and cutting tools.

Having introduced the reader to several types of drills, consideration may now be given to the various types of drill holder (or brace, etc.) which are frequently used as hand tools. There are several kinds of drill holders in common use, and each has certain advantages for various classes of work.

It is usual in engineering to refer to the drill holder, or tool in which a twist (or other type of drill) is mounted, simply as a "*drill*". Some of these tools obtain their motion by the operator merely turning a handle. Other types, although classed as hand tools, either obtain their driving power from an electric-current supply or are operated by compressed air. The latter are known as pneumatic-operated tools.

DRILLING MACHINES—HAND TYPES

Post-and-ratchet Drill (Fig. 80)

Possibly one of the earliest forms of small hand-type drilling machine to be used in engineering was the " post-and-ratchet " machine. It is still used in isolated cases, such as for outside erection work, and where no electric or other power is available.

This type of drill may also still be used for the drilling of holes of fairly large diameter, which are too large for a hand or breast drilling machine. (These are fully described later on under such headings.)

A post-and-ratchet drill consists of a round steel post mounted on a rectangular base-plate. To the post is attached an adjustable arm, as shown in the illustration. This arm can be moved up or down the post, and locked in any desired position for height by means of a fixing screw. The base-plate can be bolted to a bench or secured in position, either to a bench or to part of a work-piece, by the aid of a clamp, as is shown in Fig. 80*a*.

Positioned immediately below the sliding adjustable arm is a rather long hexagonal " pressure nut ". Attached to the lower end of the latter is a spindle which has a ratchet

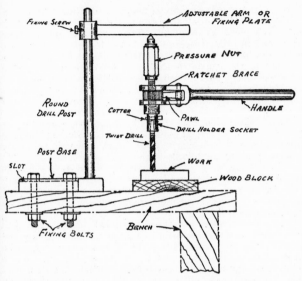

Fig. 80.—Assembly of Post-and-ratchet Drill.

(see Fig. 80b) and a handle lever mounted on it. Immediately below the ratchet device is a socket or chuck which

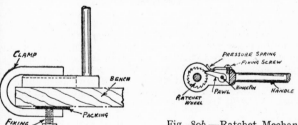

Fig. 80a.—Drill-post Clamp.

Fig. 80b.—Ratchet Mechanism (viewed from below).

forms the holder for the twist drill, or flat drill, whichever is to be used.

When the tool is assembled ready for drilling a hole, the sliding arm is first of all adjusted to the required height with the work-piece in position. The pressure nut is then tightened in order to cause the twist-drill point to engage with the work-piece. The latter should be previously marked in the centre of the desired hole position with a centre-punch. This procedure is adopted in order to assist the drill in commencing its work, and it also prevents the drill point from straying away from the correct position.

At this stage great care should be taken to ensure that the assembly is in a vertical plane. After ascertaining that the pressure nut is sufficiently tight, work can commence by simply rotating the handle of the ratchet brace.

When doing so, it is usual to turn or rotate the handle through an arc of approximately forty-five up to ninety degrees, then to return it to its starting point and repeat the stroke. The ratchet mechanism enables this to be done conveniently.

The hand strokes are thus continued, and as the drill point penetrates the work-piece more pressure is applied to the pressure nut, but it should be done gradually. The most suitable amount of pressure to be applied will soon be ascertained after a little experience. When the drill cuts into the work-piece, a little oil should occasionally be applied to the former.

As the drill cuts more deeply into the work, especially when the hole is nearing completion, there is sometimes a tendency for the drill to " bind " into the work and turn it around with the drill's rotation. This is particularly so if the work-piece is relatively small, as is shown in Fig. 80. In order to prevent the work from rotating, it may be necessary to clamp it or hold it with pliers.

Note : When the hole is nearing completion, the tendency for the work-piece to rotate with the drill is quite common, and applies to practically all drilling work.

A block of scrap wood should always be placed immediately below the work-piece in order to safeguard the bench when the drill point passes through the object which is being drilled.

It will be readily appreciated that drilling a hole by means of a post-and-ratchet drill is a somewhat slow and laborious procedure, but it is still resorted to in certain cases.

The post-and-ratchet drill can be used for drilling holes as large as one and a half inches in diameter; this would be a difficult procedure with a breast drill or a hand drilling machine of the bevel-gear type.

The clamp method of securing a post-and-ratchet drill to part of a machine is very useful, because it can often be arranged so as to clamp the post in a horizontal or an inclined position when holes have to be thus drilled.

Another advantage which is claimed for the post-and-ratchet drill is that it can often be accommodated in positions which are difficult of access for other types of manually operated drills, requiring their handles to be rotated by a circular motion.

If confined for turning space, as is often the case on repair work being done to machinery, the ratchet can often be successfully operated by turning the handle through only a small arc, say twenty degrees.

Hand-drill—Bevel-gear Type (Fig. 81)

The bevel-gear-operated hand-drill is a more modern type, and is used extensively for small work, or that of a light nature. It consists of a framework to which is attached a handle at the top end. The lower end of the frame carries a freely rotating spindle on which is mounted a small bevel pinion.

In the centre position of the framework is mounted a bevel wheel which is " meshed " or geared with the pinion. The operating handle is secured to the bevel-wheel spindle and rotates with the latter at the same speed.

Positioned at the lower end of the bevel-pinion spindle is a chuck or drill holder in which is secured the twist drill or cutting tool. The chuck is adjustable, so as to be capable of accommodating drills of several different diameters.

The bevel wheel and pinion are shown in their relative positions (in side view) in Fig. 81a. The internal mechanism of the chuck is shown in section in Fig. 81b. This comprises an outer casing, in which are mounted three sliding jaws.

The latter, being of tapered formation, move in or out from the centre, and being in three separate segments, the inner space can be increased or decreased, thus allowing different diameters of drill shanks to be accommodated. A small ball-race is usually fitted inside the sleeve on which the main spindle rotates. Fitted inside the sleeve casing are small helical springs which press the jaw segments towards the centre, as well as downwards.

To fit a drill shank into the chuck, the sleeve is unscrewed, thus opening out the jaw segments. After inserting the drill the sleeve is screwed up tightly until the segments bear securely on to the drill shank and hold it rigidly in position.

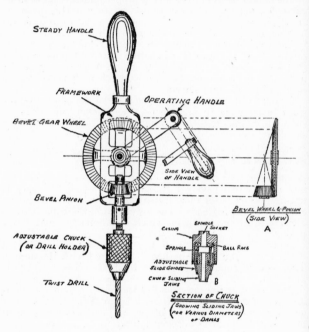

Fig. 81.—Hand-drill.

Breast-drill (Fig. 82)

In general appearance this tool resembles the hand-drill, except for slight differences in detail, and in that it is usually of larger and heavier construction.

The breast-drill is also used for larger-diameter drilling work than that of the hand-drill. One of the chief differences in the features is that the breast-drill has a steel plate at the top of the frame, whereas a hand-drill is fitted with a wooden " steady handle " at that position.

The top plate of a breast-drill is so shaped as to enable the user's chest or breast to bear on it, and from this the drilling pressure is obtained and regulated. A wooden steady handle, for gripping by the user's left hand, is situated immediately below the bevel-pinion support framework, as is shown by the illustration (Fig. 82). The operator's

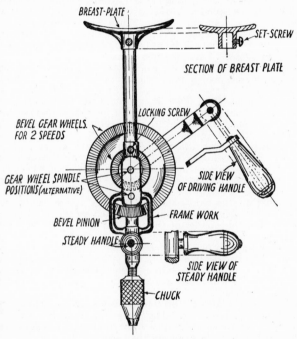

Fig. 82.—Breast-drill.

right hand is used for rotating the driving handle, which is attached to the bevel-wheel drive.

Some types of breast drills are equipped with two bevel wheels—one inside the perimeter of the other—but both are in the same plane. By operating a small screw in the frame, either of the two bevel wheels can be meshed with the bevel pinion.

The two bevel wheels are of different diameter, and each has a different number of teeth. By this arrangement, two different speeds of the main drill spindle are obtained, depending upon which bevel wheel the bevel pinion is meshed. When the pinion is meshed with the *large*-diameter bevel wheel, a high or fast drilling speed is obtained on the chuck spindle, provided that the driving handle is rotated at a uniform speed. When the bevel pinion is meshed with the *small* bevel wheel, a slow drilling spindle speed is given, if the driving handle is rotated at the same uniform speed. This principle of having a two-speed drill is often a great advantage when holes are to be drilled in certain metals which require different cutting or drilling speeds.

Perhaps at this stage it might, with advantage, be explained that if it is assumed the bevel-gear pinion has ten teeth and it is meshed (*i.e.*, its teeth are engaged) with those of the large-diameter bevel-gear-wheel which has fifty teeth, the gear ratio is said to be fifty to ten, *i.e.*, 5 to 1 ratio. This indicates that the bevel pinion revolves or rotates at five times the speed of the bevel-gear-wheel, or for each single revolution of the gear-wheel, the gear pinion makes five revolutions. Thus the gear pinion, which is mounted on the chuck spindle, revolves at five times the speed of the gear-wheel, which is rotated by the hand.

Usually, the smaller of the two gear-wheels has half the number of teeth of that of the large gear-wheel. This being so, when the gear pinion is meshed with the small gear-wheel which has twenty-five teeth, the gear ratio will then be twenty-five to ten, or $2\frac{1}{2}$ to 1 ratio. Therefore, the drill-chuck spindle will rotate at only half the speed described in the former case.

The breast-plate of a breast-drill is usually secured in its position by a set-screw. The latter can be unscrewed and the breast-plate can be adjusted by rotating it so as to give the most suitable position for the user.

Some types of breast-drills have the crank lever of the driving handle slotted at its lower end. This gives a variable length for the " lever arm ", thus enabling the user to obtain a greater " purchase " or leverage when he is turning the handle. With this type of slotted crank handle, a set-screw is provided in order to fix or locate the slot in any desired position. This arrangement will be found to be of great advantage, especially for drilling holes of fairly large diameter in a hard metal, when the greatest possible leverage is required for the driving handle.

Electric Hand-drill (Fig. 83)

In addition to the foregoing types of hand-operated drills, there are also many forms of small portable drills which obtain their rotating power from electric sources.

These are handy and serviceable tools. An electric hand-drill of the type illustrated in Fig. 83 has a small electric

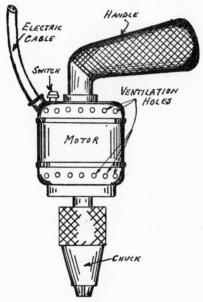

Fig. 83.—Electric Hand-drill.

motor, on which is mounted a hand-grip at one end of the motor casing. At the opposite end of this casing, a spindle projects from the motor shaft, on which is mounted the chuck or drill holder.

A flexible electric cable of suitable length is attached to the motor. The opposite end of the cable is " plugged in " to the current supply point. In a convenient position near the " hand-grip " a push-button-type switch is fitted. The whole tool can therefore be operated by one hand, thus leaving the other hand free to steady the work-piece.

This type of electric drill is very suitable for light drilling work. It is mostly used in conjunction wiith twst drills. The chuck is usually of an adjustable type, which provides for the use of several different diameters of drill.

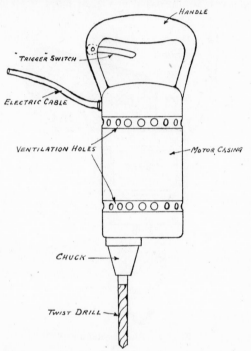

Fig. 84.—Electric Hand-drill (geared-motor type).

Electric Hand-drill (Geared-motor Type) (Fig. 84)

This drill is of heavier construction, and is therefore more suitable for work which involves the use of larger-diameter drills. It is fitted with a " trigger "-type switch, just below the handle, and the motor is of more robust construction than that which is shown in Fig. 83.

The motor, instead of having its armature shaft extended to carry the chuck, carries a gear-wheel. This gear-wheel is meshed with another, which is positioned immediately above the chuck. By this method the power of the motor is transmitted through the gears to the chuck spindle, giving it a reduced speed. This is advantageous for drilling work of a large diameter. As a precaution against grit and dirt the gear-wheels are totally enclosed inside the casing of the drill body.

Although the tool which is illustrated in Fig. 84 is of heavier construction than that described earlier, it can be operated with one hand. Some operators, however, prefer to steady the machine by holding the drill body in their left hand, whilst their right hand grips the handle and works the trigger switch.

Still heavier types of hand-operated electric drills are used for heavy-duty drilling work, but it is intended that both hands should operate and steady them, and they are designed accordingly.

Small Pneumatic Hand-drill (Fig. 85)

A pneumatic drill is operated by compressed air. It is composed of a cylindrical casing in which is housed a small air-driven motor. A pistol-grip type of handle is fitted at the top end of the casing. The air-pipe for admitting the compressed air is also positioned near the handle. Inserted in the air-pipe is a small plunger-type valve, by means of which the air can be admitted or shut off instantly.

At the lower end of the cylindrical casing the driving spindle projects, and on this is mounted the chuck or drill holder.

A pneumatic drill can be used where no electricity supply is available.

The compressed-air supply is produced in a machine called a " compresser ", and after the air has been compressed it passes to a storage cylinder. These compressers are often made in portable sizes for small classes of work.

The cylinder is known as an air receiver, where the air is stored at a suitable pressure, and from which it is piped direct to the pneumatic drill.

Pneumatic drills can also be used where, on account of the risk of sparks, an electrically operated drill would be prohibited. Such precautions have often to be taken when repair work has to be carried out in explosives factories, petroleum refineries, or any factories where inflammable gases or materials are present.

E

This type of drill is very efficient and inexpensive to operate, for several drills can be worked from one compresser unit.

The type of hand machine which is shown in Fig. 85 can be successfully used for accommodating twist drills up to about three-eighths of an inch in diameter, and is suitable for drilling work of thin character, such as sheet, plates, etc.

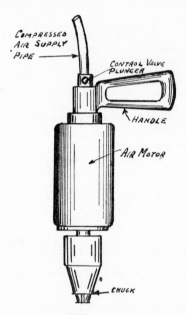

Fig. 85.—Small Pneumatic-type Hand-drill.

Pneumatic Hand-drill (Heavy-duty Type) (Fig. 86)

This drill is, in its essential principles, similar to the smaller type. It is, however, of much heavier construction, and weighs a little over twenty-one pounds. Twist drills of up to approximately one and a quarter inches in diameter can be accommodated in the heavy-duty chuck.

This drill can be used either as a portable tool or fixed in position with the aid of a post and clamp in a similar fashion to that which is illustrated with the ratchet-type drill in Fig. 80.

When used as a portable tool, the steady handle of this drill is gripped in the operator's left hand, and the "twist grip" air-valve forms the handle for the user's right hand. When so held, the weight of the tool itself often forms the

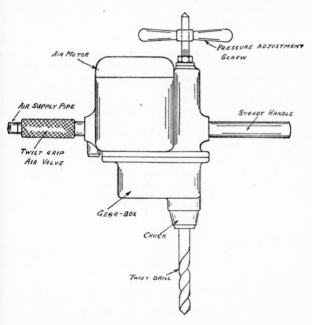

AIR MOTOR

PRESSURE ADJUSTMENT SCREW

AIR SUPPLY PIPE

STEADY HANDLE

TWIST GRIP AIR VALVE

GEAR-BOX

CHUCK

TWIST DRILL

Fig. 86.—Pneumatic Hand-drill (heavy-duty type).

pressure for the drilling operation. The air supply can easily be regulated by a twist of the right hand.

When this type of drill is to be used as a fixture, it is secured in a similar manner to that which is shown in Fig. 80. The pressure adjustment is tightened so as to engage with the arm of the "post". By so doing the reaction caused provides cutting pressure for the twist drill. If necessary, the left hand can still be used in position on the steady handle, as once the air-supply valve is opened, the right hand is free to regulate the cutting-pressure screw as the drill point advances through the work-piece.

Some drilling machines of the pattern illustrated in Fig. 86 are fitted with interchangeable-type gear-boxes, by means of which the speed of the drill-chuck spindle can be adjusted to suit different classes of work. It will be appreciated that for the drilling of holes of a large diameter the drilling speed, *i.e.*, the cutting speed of the twist drill, is much slower than that required for smaller work.

The foregoing remarks also apply to work-pieces which are very thick or of a hard nature. It is therefore an advantage to have a tool which can be fitted with an interchangeable-speed gear-box, providing it with a wider range of drilling speeds. This variation is carried out by merely changing the gear-box (which is bolted to the base of the machine) for one of a different gear ratio.

Note : Bench-type drilling machines and heavy, power-driven types are dealt with in Volume III.

Hand-reamer (or " Rimer ") (Figs. 87, 88, and 89)

A reamer is a tool which is used in order to enlarge a previously drilled hole. This tool is also sometimes called a rimer. It will be appreciated that after a drill has been in constant use for a considerable period, its side walls or side cutting edges wear down. Holes which are drilled by such a

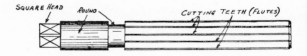

Fig. 87.—Straight-flute-type Reamer.

tool will therefore be slightly undersized. In such cases a reamer can be used to cut out a very small amount of metal to bring the drilled-hole diameter up to its correct size.

It is made of very hard steel, and its cutting edges are very sharp, so that after use, a clean, neat finish is produced on a hole.

A hand-reamer is used in conjunction with a wrench or handle. The handle is exactly like that which is used for screw taps, as is illustrated in Fig. 64*b*, for a reamer has a square top end similar to a screw tap.

There are several types of reamers in use, and some of the most common are shown in the illustrations. A reamer somewhat resembles a twist drill, for it has similar-shaped cutting edges, though sometimes the latter are perfectly straight. Those of the spiral type, shown in Fig. 88, do

Fig. 87a.—Section of Reamer.

not advance with so rapid a helical formation as that of a twist drill.

A typical cross-section of a reamer is shown in Fig. 87a. On reference being made to the figure, it will be seen that the cutting edges taper slightly, thus forming a clearance.

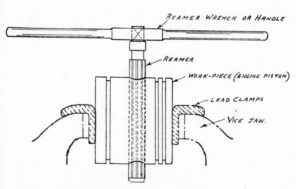

Fig. 87b.—Application of Reamer.

The " flutes " or spaces between the cutting teeth form recesses into which the cuttings collect.

It will be seen that types which are illustrated in Figs. 87 and 88, have parallel sides, therefore these types are used for " reamering " out or enlarging similar-shaped holes. The

type shown in Fig. 89 is used for enlarging tapered holes.
The lower ends of all reamers taper slightly as a means of
better enabling them to enter the holes concerned.

A modern type of reamer is now made, which is of the
expanding kind, the cutting blades of which can be adjusted

Fig. 88.—Spiral-flute-type Reamer.

by manipulating a fine screw-thread, which increases or
decreases the overall diameter slightly. This type is now
extensively used in large machine shops.

A reamer is usually used for fine work or in cases where a
very accurate fit is desired, such as on the connecting-rod of
a steam-engine or the " gudgeon-pin " holes of a piston for
a motor engine.

Fig. 89.—Tapered Flute-type Reamer.

When using tapered reamers, great care must be taken to
ensure that even pressure is applied so as not to cut away too
much metal from one side of the hole. The depth of
penetration must also be closely watched, so as to produce
the correct taper.

A typical use of a parallel-sided reamer as applied to the
gudgeon-pin holes of a motor piston is shown in Fig. 87b.

CHAPTER V

PIPE-WORK TOOLS

QUITE a large number of tools which are primarily used in conjunction with pipes or tubes can also be used for handling round bars, etc. Certain tools of a special character are used for the cutting of large pipes. Small pipes or tubes can often be cut efficiently by the use of a hack-saw.

However, for pipes of larger diameters, special " pipe-cutters " are used. By means of these, a clean cut can be effected fairly quickly, leaving a true square edge, which is ready for the application of any screw-threading apparatus.

A special type of vice is required in which to secure large-diameter pipes on which any work, such as cutting, screwing, or bending, has to be performed.

In order to screw-thread pipes externally, stocks and dies are used. These tools have already been fully described in Chapter III, under the relative heading of " Screw-threading Tools ". In the same chapter internal screw-threading tools, called screw taps, which may be used for pipe-work, were also dealt with.

Further tools applicable to pipe-work will now be described.

PIPE-CUTTER (FOUR-ROLLER TYPE) (Fig. 90)

A pipe-cutter consists essentially of a strong frame which usually takes the form of a large hook or letter " C ". Projecting at one end of the frame is a screwed spindle, which also forms the handle of the tool, as is shown in Fig. 90. Two pairs of guide rollers are mounted at the opposite end of this screwed spindle, and these rollers are arranged so as to rotate freely on their hinge pins (see Fig. 90a). One set of rollers is positioned immediately above the other and forms a guide as the tool rotates around the pipe. At positions on the " C " frame immediately opposite those of the guide rollers, two specially hardened " cutter-wheels " are fitted. These also rotate freely on their hinge pins (see Fig. 90b).

In order to use a pipe cutter, the pipe should be secured in a horizontal position in a vice. The pipe-cutter spindle handle is then unscrewed sufficiently to allow the work-piece (the pipe) to be inserted between the guide rollers and the cutter wheels, as shown in Fig. 90. The handle is then screwed up in order to form a fairly tight grip between the

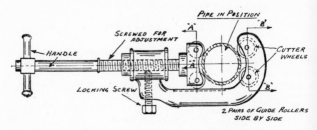

Fig. 90.—Pipe-cutter (four-roller type).

cutter edges and those of the rollers. Next, the whole tool is rotated bodily around the pipe, and the pressure of the grip causes the cutter wheels to commence cutting the pipe. After one, or possibly two, complete revolutions have been

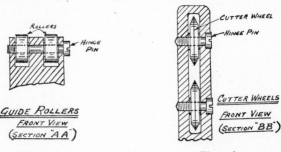

Fig. 90a.

Fig. 90b.

made by the tool around the pipe, further pressure is applied by screwing up the spindle handle. More revolutions of the tool are then made and pressure is increased occasionally until the pipe is effectively cut through. It is beneficial, when more pressure is added, to occasionally apply a little oil. This assists the cutting action in a similar way to that applied during screw-threading when stocks and dies are being employed.

Before a pipe-cutter is applied to any work the position where the pipe is to be cut must be marked exactly, so that when the cutter wheels are placed in position they can then be lined up with the mark.

After a pipe has been cut it will be found to have a slight taper at its end. Sometimes the ends will have bulged out slightly, or expanded, at the top of the taper. This bulge can easily be reduced by filing it down. The tapers and bulges on the two ends of the cut pipe are, of course, due to the " Vee " shape of the cutter wheels.

As the purpose of cutting a pipe is usually to make it of suitable length to form part of some installation, it will probably have to be screw-threaded. When filing down the bulges, it is also advantageous in such cases to form a gradual

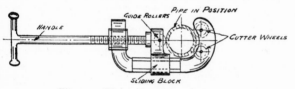

Fig. 91.—Pipe-cutter (two-roller type).

taper by filing the pipe end for about half the length of the required threaded section. This greatly assists in the screw-thread formation, for when the stocks and dies are applied it forms a " lead " for the die teeth to commence their screw-cutting action.

The type of pipe-cutter which is shown in Fig. 90 is suitable for pipes of approximately three inches up to six inches in diameter. Suitable pipe-cutters can be obtained for each range of pipe sizes likely to be encountered.

PIPE-CUTTER (TWO-ROLLER TYPE) (Fig. 91)

The two-roller type of pipe-cutter is in general respects similar to the four-roller type. It has only two guide rollers, instead of four, and is not of quite such heavy construction. There is also a sliding block fitted to the frame in which the guide rollers are positioned, as is shown in Fig. 91. This tool is used for pipes of approximately one and a half up to three inches in diameter.

PIPE-WRENCHES (ADJUSTABLE TYPES)

In order to hold pipes securely or prevent them from rotating, a special tool is used. For pipes of small diameter a pipe-wrench is often sufficient for this purpose. A pipe-wrench, which is similar in shape and general construction to the tool illustrated in Fig. 7a, may be used. The only real difference between this and an adjustable spanner is that a pipe-wrench has a series of " saw teeth " on each jaw face in order to enable it to grip well into the metal of the pipe. As an alternative, a tool similar to that shown in Fig. 47 may be used.

These tools are frequently used for tightening up unions or pipe sockets which form pipe connections. In such cases the length of pipe concerned may be secured in a pipe vice while the socket is fitted.

On other occasions the socket or union might have to be fitted to a pipe which is already fixed in some installation or circuit. In this case two pipe wrenches would probably have to be used ; one for preventing the pipe from rotating, and the other for fixing the union.

These tools are very suitable for pipes of two or three inches in diameter. For pipes of small diameter, say those up to three-quarters of an inch, gas pliers or a small-size wrench would be quite effective.

CHAIN PIPE-TONGS (OR "CHAIN-DOGS") (Fig. 92)

" Chain-dogs " are used to secure large-diameter pipes or to prevent them from turning round. Chain-dogs are also used for tightening up the unions of large pipes.

Chain-dogs or chain pipe-tongs consist of a substantial steel handle of round-bar form on which is mounted a specially shaped lug. This lug is of U-shaped formation as shown in the plan view of Fig. 92. Both arms of the U are pear shaped and have teeth formations on them.

Inserted between the arms of the U is another lug or boss which has larger teeth on its upper and lower faces. These teeth are suitably arranged to accommodate the links of the chain, as shown in the side view of the illustration, Fig. 92. A suitable length of strong, roller-type chain is pivoted and secured by one end of the U-piece by a stout pin.

To use the chain-dogs, the jaw of the U-lug is placed over the pipe, and the loose end of the chain " lapped " around the pipe's circumference. The chain links are pulled reasonably tight over the pipe and secured in the special teeth of the centre lug, as shown in the illustration. The handle of the tool is then lifted upwards, causing the teeth of the U-lug to grip the pipe. The handle can be manipulated, by pushing it slightly forward, as well as upward, until a secure grip is obtained. Once the pipe is gripped, the whole tool and the pipe can be rotated.

The chain-dog is a substantially built tool, and will stand up well to any amount of hard use. The only disadvantage of this tool is that its teeth mark the outer face of the pipe. These marks can be erased by gentle filing once the pipe or union is finally screwed up.

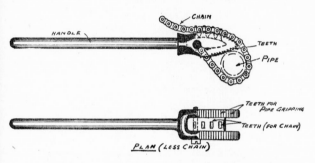

Fig. 92.—Chain Pipe-tongs (or " chain-dogs ").

After much use the teeth of the chain-dog get clogged with the small bits of the " pipe marks " and should be cleaned out occasionally, otherwise the tool may tend to slip instead of gripping the pipe. If necessary, chain-dogs can be used by moving them through an arc instead of a complete circular motion. The grip is then released for each turn or arc and a fresh grip made on the pipe for the next turn.

PIPE-VICE (Fig. 93)

This vice is specially adapted to secure pipes or round bars while work is performed upon them. It will be seen, on reference being made to Fig. 93, that the framework is hinged or split in halves. This is to provide for lengths of pipe which may have "elbows" or T-sockets, mounted on their ends, and which could not easily be passed between the jaws.

In such cases, the catch can be lifted or released, the top jaw raised and the pipe laid in position. The top jaw can be replaced and again clamped in the fixed position by the catch.

A pipe-vice, if of a light type, is often mounted on a light

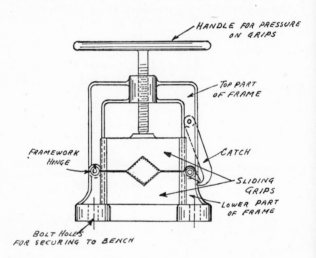

Fig. 93.—Pipe-vice.

triangular frame or floor stand. Heavier types of pipe-vices are bolted down to substantial steel or wooden benches.

The inner faces of the vertical frame sides are fitted with guides. Into these guides the jaws are positioned, and the space between the jaw faces is regulated by a screw spindle which has a handle at its top end. The jaw faces are "Vee"-shaped, and have teeth on them to enable the pipe to be effectively gripped.

A pipe-vice will usually accommodate several different sizes of pipes, and various sizes can be obtained, which cater for a whole range of assorted pipe diameters.

PIPE- OR BAR-BENDER (Fig. 93a)

In order to bend pipes or bars of comparatively small diameters, the tool shown in Fig. 93a may be used. This tool can be used successfully for bending pipes when they are cold. Larger pipes or bars frequently have to be heated in order to bend them more easily.

The tool illustrated is an elementary form which is

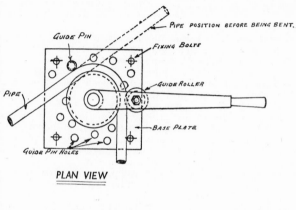

PLAN VIEW

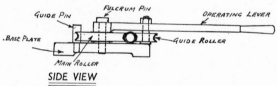

SIDE VIEW

Fig. 93a.—Pipe- or Bar-bender.

extensively used. It consists of a base-plate which is made of strong steel plate or cast iron. There are holes provided in it near each corner for bolting it securely to a strong bench or stand. In the centre of the base-plate is a strong steel fulcrum pin, on which is mounted a loose roller, and attached to the fulcrum pin is a stout steel lever which is fitted with a guide roller and pin as shown in the illustration.

The base-plate is also provided with holes at various radii

from its centre. These holes are positioned to provide a suitable place in which to fix the guide-pin.

Several rollers, each of different diameter, are provided in order to cater for pipes or bars of varying sizes, and to provide for the formation of bends of various sizes. By merely removing the fulcrum pins the rollers can be taken out and replaced by those of another size.

In order to use a pipe-bender, a suitable size of roller is fixed to the base-plate and secured by the fulcrum pin. A suitable-sized guide roller is also fitted to the lever, and secured by its pin.

The straight pipe is then placed in position into the groove of the main roller. The guide-pin is placed in whichever hole of the base-plate is found most suitable for it to press up against the side of the pipe, in order to fix it in position. The hand-lever carrying the guide roller is then positioned close up to the pipe, near the guide-pin. (*Note :* At this stage the pipe is, of course, straight.)

The hand-lever is then gripped and pulled in a clockwise direction in order to bring it into the position shown in the illustration. This pulling action causes the pipe to traverse the path swept by the guide roller of the lever to which the roller is attached, and the pipe is thus bent.

The amount of bend in the pipe depends, of course, on how far the lever is pulled round, or on how large an arc is swept through by the hand pull.

The thickness of the pipe walls, and the material of which the pipe is made, also have a bearing on the size of bend which can be thus effected. The diameter of the main roller is also a deciding factor for settling the size of radius to which the pipe can be bent without unduly stressing or damaging the pipe. Some classes of pipes can be bent more easily than others. On some occasions, pipes of a delicate nature, such as those which are made of thin brass or copper, are filled with molten lead before attempting to bend them. After the molten lead has solidified the pipe becomes equal to a solid bar and can be bent more easily, and often without damage being done to it. After bending the pipe by this means, it is heated, the lead melted and extracted or " run off ", as it is termed.

Sometimes, in order to effect a " slow bend " (*i.e.*, one of a small arc formation) a pipe is filled with sand, which is tightly rammed into it, to form a more solid mass of material. It can then often be bent while it is cold, and after bending, the pipe is gently tapped with a hammer to shake out the used sand. When using a pipe-bender the lever handle can be gripped by either one hand or both, depending upon the amount of effort required and the size of pipe concerned.

Several other forms of pipe-benders are used in engineering, some of which are fitted with a worm and worm-wheel, through which the pressure is applied to the lever for giving the bending action. These types can be used for much larger pipes than the tool which is illustrated.

For the bending of very large pipes, machines which are hydraulically operated are used extensively, as they can exert tremendous pressures.

The type of bending tool shown in the illustration is extensively used for bending small reinforcing bars for reinforced-concrete work, as they can easily be bent cold.

CHAPTER VI

LIFTING TOOLS

SCREW JACK (HEAVY INDUSTRIAL TYPE) (Fig. 94)

ONE of the commonest forms of lifting tools is, perhaps, the
" screw jack ". It is shown in elementary form in Fig. 94,
and this type is sometimes called a " bottle jack ", possibly
because its general shape somewhat resembles a bottle.

This tool consists of a strong cast iron, or steel, body, into
which is screwed a substantial square-threaded steel screw.
At the top of the screw is a boss or head. The
latter is of massive construction and has holes through
which passes a steel handle or large type tommy bar.
From the top of the screw head a lug projects, which forms
a pad. The pad, however, is loosely attached to the boss,
either by a sliding peg or a set-screw, in order to allow the
pad to move upwards in a vertical plane, while the screw
boss is rotated. The top of the pad is provided with small,
pyramid-shaped teeth, which grip the work-piece and
prevent any slip occurring.

When the screw is in its lowest position it rests inside the
jack body, for the top part of the body is internally screw-
threaded and thus acts as a large nut. If the jack (with its
screw positioned well into the body of the tool) is placed
beneath any large or heavy object, the latter can be lifted
by rotating the screw so that it moves upwards. As most
screw jacks are provided with right-handed screw-threads,
the rotation would be in an anticlockwise direction.

Some types of screw jacks are provided with two screws—
one inside the other. By these means, double the vertical
lift can be obtained, for the inner screw is first extended,
then, as it projects to its full extent, the outer one can be
unscrewed. With these types of screw jacks, loads of two
or three tons can be lifted.

A modification of the last type is one where, instead of the
handle being used directly with the screw, a worm-wheel
and pinion are meshed with it, and the handle is
applied to the worm-wheel. With this arrangement the
pinion is loosely mounted on to the screw. The latter

passes through the centre of the pinion, which acts as a nut. When the pinion is rotated, by being meshed with the worm-wheel, and as both are in relative fixed positions, the screw moves up or down through the pinion.

By this arrangement, a great purchase is obtained through the worm-and-pinion gearing, so that greater loads can be lifted.

MOTOR-CAR-TYPE JACK (Fig. 95)

This type of jack is of much lighter construction than those which have been previously described. It has a

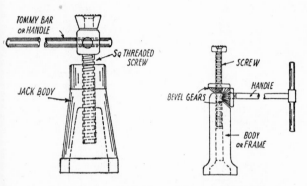

Fig. 94.—Screw Jack or " Bottle Jack." Fig. 95.—Motor-car-type Jack (light type).

lighter type of body and a square-threaded screw, but is fitted with a head pad similar to the foregoing types.

With this type of jack the operation of the screw is effected by bevel gears, which, although free to rotate, are located in fixed positions. As will be seen, on referring to the illustration, the screw passes through the top bevel wheel.

The lower bevel wheel is meshed with the top one, but the former has a square recess, into which the handle lever is fitted. On the handle being rotated, the bevel wheels also rotate, but they do not move bodily. Owing to the screw passing through the top bevel wheel, which is also internally screw-threaded to suit the screw, the latter is extended so that any object immediately above the screw pad is lifted.

The car jack is capable of lifting weights up to about one ton, which is the average weight of a medium-sized motor car.

HYDRAULIC JACK (Fig. 96)

A hydraulic jack is one which is operated by some liquid, usually oil. The principle on which this type of jack works is that oil is forced into a cylinder. Inside the cylinder is a " ram " which has a lifting pad at its outer end. When oil, which cannot be compressed, is pumped into the cylinder under great pressure, the ram must move to make room for more incoming oil. In so doing it lifts the load.

In order to stand up to the work which they are called upon to perform, hydraulic jacks must be made of tough material. Gunmetal and steel are, therefore, frequently employed in their construction.

The elementary form of such a jack is illustrated in Fig. 96. It comprises a frame (which is recessed so as to form an oil reservoir), a cylinder, and a ram. The oil reservoir, which also forms the jack base, is fitted with an oil-filler hole and a plug. Down the centre of the reservoir passes a tube, at the lower end of which is a non-return valve through which the oil enters from the reservoir. This admission of the oil is effected by the plunger being lowered by manipulating the lever. Suction causes this valve to open and oil is admitted to the central tube until it is full.

The operating lever is then pressed down, causing the plunger to rise in the tube. The non-return valve at the top of the tube is opened by the pressure of the oil, and the oil is driven into the cylinder. The valve then closes, thus preventing any oil from returning to the tube. The oil, on entering the cylinder, forces the ram upwards, and, in consequence, it lifts the load on the pad at the ram top.

If the handle of the lever is raised, the plunger is lowered, and this tends to create a partial vacuum in the central tube. In so doing the suction again opens the inlet non-return valve at the tube base. More oil is admitted through the valve from the oil reservoir. On pressing down the lever handle again, the plunger rises and drives the oil up the tube, through the non-return valve at the tube top, into the cylinder. As this oil enters the latter the ram is forced higher up the cylinder, thus raising the load on the ram pad still higher.

By simply manipulating the hand lever up and down, the process of admitting more and more oil through the valves into the ram cylinder is continued until the load has been lifted to the desired height. In order to lower the load, the " lowering " valve is opened, and the oil flows quickly from the cylinder back into the oil reservoir, through the small hole provided for that purpose, and the ram is lowered.

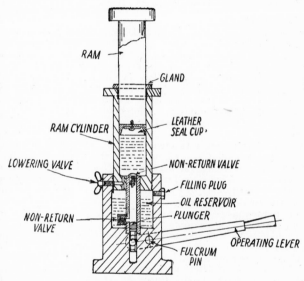

RAM

GLAND

RAM CYLINDER

LEATHER SEAL CUP

LOWERING VALVE

NON-RETURN VALVE

FILLING PLUG

OIL RESERVOIR

NON-RETURN VALVE

PLUNGER

FULCRUM PIN

OPERATING LEVER

Fig. 96.—Hydraulic Lifting Jack.

The speed of lowering the load is governed by the amount by which the lowering valve is opened. This can be regulated as necessary.

For the efficient operation of a hydraulic ram it is essential to keep all joints and valves absolutely leak-proof. Some types of hydraulic jacks are fitted with extra refinements, such as air-release valves, which are provided so as to allow the air to escape when filling the reservoir with oil. After filling this with oil, both the air-valve and the filler plug must be screwed up tightly and kept leak-proof.

The leather " seal cup " at the ram base is so arranged that as oil is admitted under pressure the walls of the cup are forced against the cylinder walls, thus forming an efficient seal. At the top of the cylinder some jacks have an extra seal in the form of a gland which can be packed with oiled twine or similar material, and thus prevent leakage.

CHAIN BLOCKS (Fig. 97)

For overhead lifting of very heavy loads, cranes of various types are used, but for lifting relatively small loads, chain blocks are frequently used.

In their simplest form, chain blocks consist of two pulleys which are fitted side by side. These pulleys are recessed on their perimeters to accommodate a chain or wire rope. The pulleys have diameters of different sizes, and they are mounted on one common spindle, which is slung in a frame. At the top of the frame is a " sling hook " which is used for securing the pulley blocks to any suitable overhead structure.

Immediately below the two pulleys is a third, which has a spindle mounted in a frame. This frame has a " lifting hook " to which is attached the load. Lapped around the lower pulley is an open-linked chain (or wire rope). This chain passes thence to the groove of the top pulley which has the smaller diameter.

It is then lapped around its perimeter, within the groove, and passes in " open sling " form, as a " hand-grip " loop of chain. It then continues around the large-diameter pulley groove, and thence to the lower pulley again, as is shown in the illustration.

The top pulleys having different diameters, their radii are also proportionately different. If the lifting hook is attached to a load and the fixing hook is fastened to any overhead structure, the load can be lifted by pulling on the " free " loop of the chain, which passes over the larger of the top pulleys.

The lifting ability is thus accomplished because the larger pulley, which has a larger radius, gives a greater " leverage " over the radius of the smaller pulley. The top pulley frame of some " blocks " is sometimes fitted with " guides " which contain " catches " for preventing the load from " running back " between each individual hand-pull.

By this arrangement, the load can be lifted off the ground

and retained in any position, even when the hand-grip is released from the chain.

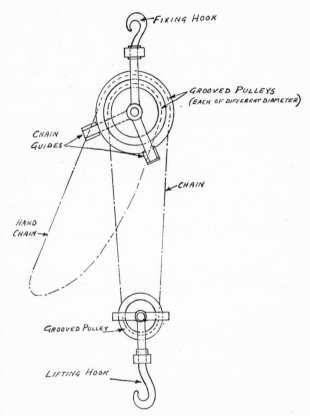

Fig. 97.—Chain Blocks.

This type of pulley block is called a " Weston block " after its originator. It is frequently used in engineering works for lifting heavy machinery, in the absence of an overhead travelling crane, and in garages, for lifting the engines of motor cars from their chassis, etc.

This type of pulley block is also used extensively for outside erection work in conjunction with a " pole " and guy ropes, for lifting loads. Sometimes it is slung from the top joint of three poles, whose legs are splayed out in triangular form over the work-piece to be lifted.

The pulleys are sometimes referred to as " sheaves "; and other types are sometimes used which, instead of having two pulleys (or sheaves) at the top, are fitted with several, each of different diameter.

The rope is wound over one size, then over the next smaller size pulley, and so on in sequence. One end of the loop thus passes over the largest pulley, and the other end over the smallest pulley. With this arrangement, still greater leverage is obtained, because the hand-pull is exerted on the free end of the rope sling from the largest pulley, and the " paying-out " length of the rope sling passes to the pulley of the smallest diameter.

Another form of pulley block is one which has a worm and worm-wheel application, by means of which a large wheel is fitted to the worm shaft, on which the hand-pull is exerted. A greater leverage or greater purchase is obtained by this arrangement. With this type of block, an automatic ratchet-type catch is fitted in order to provide against any run-back of the load when the hand-pull is released.

CHAPTER VII

FORGING TOOLS

" FORGING " is the term applied to " work " which is " beaten ", " forged ", or " hammered out " to a certain required shape while it is still hot.

Although hand forging has of recent years been superseded by machine forging and " drop-stamping " or " pressing-work ", it has to be resorted to for certain classes of work. If an occasion arises when a round steel bar has to be " cranked " or bent to form a curve, and if it is too large to be bent cold, it must be heated and then hammered to the desired shape. This is called forging.

Furthermore, special bolts such as in the case of an " eye "-bolt or a long foundation bolt have to be hand-forged.

In order to carry out these operations, certain tools must be used. These are called blacksmith's tools, for a blacksmith is one who works on the forging of metals in their " hot " state.

One of the principal tools is a " forge " or " hearth ". This is usually an open coke fire to which is applied a blast of air.

Greater heat is generated by this method than can be obtained from a coal fire. Coke is also much cleaner in use than coal, which contains oils and tar substances. A popular type of small forge suitable for hand-forging work is one which comprises a flat, cylindrical steel tray or hearth of approximately three feet diameter with sides about four inches deep.

This is mounted on a steel stand of a convenient height for hand work. Immediately below the hearth is a system of air " bellows " which is operated either by hand or by foot. A pipe connects the bellows and the hearth. At the rear of the hearth side or wall, where the air pipe enters it, is a nozzle or jet which is placed so as to feed air horizontally towards the hearth centre.

Coke is placed on the tray, and a fire lighted by the addition of bits of wood or oily rag, etc. Air is then introduced by operating the bellows until the coke is red hot.

At this stage the forge or hearth is ready for heating the work-piece. The latter is placed in the centre of the hearth, and the hot coke is arranged almost to cover it. The bellows are then operated to cause a constant air blast, by means of which the work-piece is finally heated.

Incidentally, the above-mentioned type of hand-forge is frequently used for " hotting-up " rivets which are used for hot-riveting work. It is used extensively for this purpose in shipyards.

Improved types of hearths sometimes have small electric fans fitted in order to generate the air blast, instead of the hand-operated air bellows. The principle of all forging work is to heat the metals up to a suitable fusion point and then hammer them together so as to form one unit. If two

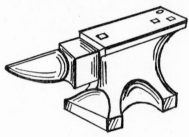

Fig. 98.—Anvil.

pieces of metal are so joined together they are said to be " welded ".

ANVIL (Fig. 98)

This tool, usually made of solid wrought iron or cast steel, forms the " bench " or " block " on which small forgings are hammered into shape. The top part of an anvil is mostly made of tough steel in order to stand up well to its duty.

Its shape is shown in Fig. 98, from which it will be seen that it has a flat top which contains some square and some round-shaped holes. These are used for the insertion of one end of the work-piece during the bending or shaping of the forging. The holes are also used for inserting allied tools, such as " swages ", " cutters ", etc., during the forging processes.

From the illustration, it will also be seen that the anvil has a substantial pointed projection from a square shoulder, at one end of the top piece. These formations are used for bending and shaping the work on.

An anvil is of robust construction, and may be two and a half or three feet in length. It may even weigh two or three hundredweights. The " French " type of anvil has a long, pointed projection at each end of the top piece.

SWAGE BLOCK (Fig. 99)

A swage block is a solid, rectangular block of either forged or cast steel. It has assorted shaped recesses on all its

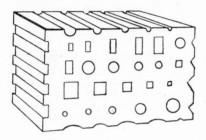

Fig. 99.—Swage Block.

faces, and various shaped holes or perforations through its sides. This block, like the anvil, is used for shaping various work-pieces. It is also used for inserting in its hollows and holes various allied tools, such as " swages " of different kinds.

A swage block is generally used for the " finishing off " of a work-piece. If the latter, after preliminary forming on an anvil, is of a round bar shape, it is reheated, and transferred to a similar-shaped recess of a swage block. An allied tool of similar shape, called a swage, is then placed on top of the work and struck by a hammer in order to complete the shape.

Although of robust construction, a swage block can be turned about so that any of its faces is in the upward position, consequently any face which is found to be most suitable for the work-piece concerned can be uppermost.

BLACKSMITH'S TONGS (Fig. 100)

These are somewhat similar to a large pair of pliers, except that the handles are much longer, on account of the heat of the hearth, from which the work-piece has to be extracted. Tongs are also used to hold the work-piece in their jaws during the forging process.

When a grip is made on the work-piece—while the latter is in the hearth-fire—a steel ring is slid over the tapered handles of the tongs. It is pushed along them until it is tight, in order to effectively secure the work between the tong jaws, by the reaction of the handle grip.

Although the tongs which are shown in the illustration have flat jaws, they may have other shaped jaws, either of a hollow curved formation or hollow square recessed shape. The different jaw shapes are used to suit the nature of the work-piece.

Tongs are usually made of wrought iron or mild steel.

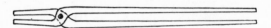

Fig. 100.—Blacksmith's Tongs.

The jaw sizes naturally depend on the class of work for which they are intended.

HOT SETT (Fig. 101)

This tool is used for the cutting of hot metal during the forging process. It is to a blacksmith just what a cold chisel is to a fitter, and it is used for a relatively similar purpose, *i.e.*, cutting the metal. A hot sett is made of hard steel, but it is of larger diameter and of a shorter length than a cold chisel.

Its point is not so sharp, and instead of it being held directly in the hand, it has a long handle. The handle is usually formed by twisting stout wire or thin steel rod once or twice around the sett's shank, the ends projecting for about three feet or so, as shown in Fig. 101. Some types of handles are made by twisting thin canes or hazel twigs around the shanks, for these do not jar the holder's hand as does a steel rod.

The sett is held by the blacksmith's left hand, while he strikes it with a hand-hammer, using the latter in his right hand. If, however, the work-piece is large, the blacksmith holds the sett, and his " mate " (the " striker ") uses a sledge hammer for the striking of the hammer blows. In this case the blacksmith would steady the work by gripping it with tongs in one hand, and the sett would be held in his other hand.

STRIKING HEAD

CUTTING EDGE.

Fig. 101.—Hot Sett.

When a blacksmith's striker, *i.e.*, his mate, is using a sledge hammer he generally works left handed (*i.e.*, he swings the hammer over his left shoulder) almost facing the blacksmith. This is advantageous because, in this position, the hammer is farthest away from the smith's normal working position.

Also, by adopting these positions, both the smith and his striker have unobstructed views of the whole of the work-piece, whereas if the striker stood at the smith's side his view would be partially obstructed during the swinging action of his body, whilst delivering the hammer blows.

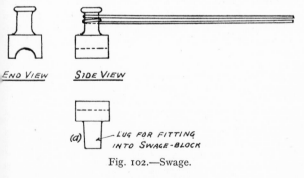

END VIEW   SIDE VIEW

(a)   LUG FOR FITTING INTO SWAGE-BLOCK

Fig. 102.—Swage.

SWAGES (Fig. 102)

A swage is used by the blacksmith for the shaping of cylindrical objects. It is made of tough steel, to a formation similar to that shown in Fig. 102. The blacksmith holds the work by tongs in his left hand, and the swage in his right hand over the work. His striker uses the hammer.

In conjunction with the above-mentioned tool, a swage similar to that depicted in Fig. 102a is used, fitted to a swage block. Some such swages have round-shaped lugs, square lugs, or other shapes.

Other types of swages are used for special work, but those shown in the illustrations are some of the commonest forms used.

CHAPTER VIII

RIVETING TOOLS

FOR hand-riveting work of a small character, the " ball-pein "-headed hammer is used, similar to that described in Chapter I, and shown in Fig. 15. Such a tool is often used for cold riveting work; but for rivets of larger diameter, and when they are used hot, a heavier type of hammer is used in conjunction with various other tools.

In all cases of riveting work, the rivet (after it has been placed into position in the work) must have its head well supported on a solid block or a cup-shaped tool into which the head can be fitted and be well supported. The projecting shank of the rivet is hammered to a rough formation of its head. A cup-headed tool is then applied—before the rivet cools off—and the tool is hammered in order to complete the head formation.

For work of a special nature, where a watertight joint is essential the rivet heads and also the joints in the plates are then sealed more effectively by what is known as " caulking " and " fullering ". Special tools are used for these purposes. If conical-shaped rivet heads are desired, special tools must be used in order to give this type of head.

If the rivets are to be countersunk, the rivet holes must be previously prepared, and this involves the drilling of the countersinks, etc.

Hammers of various weights are used for riveting, each depending on the class of work and the size of the rivets to be used. In modern methods, for medium-sized riveting which involves a large number of rivets being used, a tool known as a pneumatic hammer is extensively employed.

HAND RIVETING HAMMER (Fig. 103)

This is a medium-sized hammer, between the size of a hand hammer and a sledge type. Its weight is approximately three and a half pounds.

It is used for hot-riveting work, in order roughly to form the head of a rivet prior to finally shaping it with a " cupping-tool ". This type of hammer is used for the

F   157

formation of conical-shaped rivet heads. It may also be used for the formation of " semi-countersunk " heads, or the types of rivets which are fitted to countersunk holes,

Fig. 103.—Hand Riveting Hammer.

but whose heads are roughly flattened, and which project slightly above the surface of the work, so that they are not entirely countersunk. This hammer is held in both hands, and not like the hand hammer, which is held in one hand only.

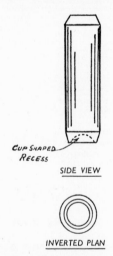

SIDE VIEW

INVERTED PLAN

Fig. 104.—Hand Snap (or " Dolly ").

The riveting hammer is lifted a certain amount, and then brought down quickly in order to strike the blow. It is not used like a sledge hammer, which is " swung " with a complete circular motion.

HAND SNAP, DOLLY (OR CUP TOOL) (Fig. 104)

Note : The term " snap " is most commonly used, but
in some parts of the country it is referred to as a " dolly ".

A " snap ", " dolly " or " cup-tool " is used for completing
the rivet-head formation after it has received a preliminary
hammering to form its rough shape. A hand snap is
made of hard steel, and has a cup-shaped recess at its lower
end. The snap is placed over the partly formed rivet
head, and is firmly gripped in the left hand. A hand
hammer, which should be held in the right hand, is then
used—face downwards—for striking the snap. The
latter, after it has been used on the rivet in a vertical plane,
is slightly tilted so that the edges of one side of the cup
formation completely engage with the edge of the rivet head.
The snap is then hammered, and moved all around the
rivet head until the edges of the latter have been entirely
encircled, and the head is well formed. It thus grips the
face of the work so as to form a seal.

For other types of rivet heads, such as " conical " or
" pan "-head rivets, snaps with those shaped recesses
must, of course, be used.

Note : The various types of rivet heads are fully dealt
with in Volume II (Components) under their respective
headings, but this volume is intended to introduce to the
reader the various *tools* which are used in engineering,
together with their uses, and *not* the " components " upon
which they are used.

LARGE SNAP, DOLLY (OR CUP TOOL) (Fig. 105)

This type of snap is used for larger sizes of riveting work.
It is therefore provided with a handle for holding by one
man or boy, while it is hammered by another operator.

Some types have handles which are formed by twisting a
piece of thin steel rod around the shank, as shown in the
illustration. A disadvantage of this type of handle is that
the holder suffers from the jarring action which results
from the hammer blows, unless leather or rubber gloves are
worn or canvas is lapped around the steel handles. Other
types of handles for this class of snap are often formed by
providing a hole through the shank, and inserting a wooden
shaft, similar to a sledge hammer.

In addition, this type of snap is also used as a bolster or
support for the formed rivet head, whilst the other head is

being formed on the rivet's blank end. When so used it is wedged between the work and a solid block during the formation of the hot rivet head. Although Fig. 105 shows a cup-shaped recess, other types are frequently used for different shaped rivet-head formations.

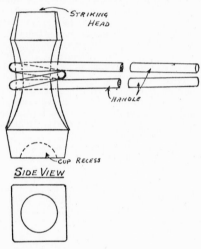

SIDE VIEW

PLAN OF CUP END

Fig. 105.—Large " Snap ", " Dolly " or "Cup-tool ".

CAULKING TOOL (Fig. 106)

This tool somewhat resembles a cold chisel, except that its blade and point are slightly different in shape. The tool is made of hardened steel, and it is used for sealing more effectively the plate edges of a riveted joint. In some cases it is also used on the edges of a rivet head to obtain a better seal between the head and the plate to which it is riveted.

The blade point of a caulking tool is not so sharp as that of a cold chisel, but is blunted so as not to cut the rivet head, but just to burr it slightly.

In order to caulk the joint between two plates which are riveted together, the edge of the plate which forms the

" lap " should be driven tightly into contact with the face of the other plate. This is effected by using the caulking tool as a chisel, and hammering it with a hand hammer in a similar way to that used for " chipping " work.

The action is somewhat similar for caulking a rivet head, but the tool is gradually worked all around the rivet head during the hammering process.

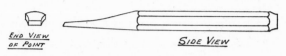

END VIEW
OF POINT

SIDE VIEW

Fig. 106.—Caulking Tool.

FULLERING TOOL (Fig. 107)

This tool is similar to, but of more robust construction than, a caulking tool. It has a wider flat face and is not so pointed as the latter.

A " fullering " tool is not used on rivets, but only for sealing the edges of plates. The latter are usually pre-

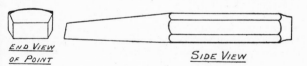

END VIEW
OF POINT

SIDE VIEW

Fig. 107.—Fullering Tool.

viously prepared by " chamfering " or tapering the edges before the plates are riveted together.

Note : The application of this process is dealt with more fully in Volume II, under " Rivets and Riveting ".

HAND-TYPE PNEUMATIC RIVETING TOOL (Fig. 108)

This tool is used for hot riveting. The type illustrated has an adjustable snap-holder permitting the use of various sized snaps suitable for rivets up to 1 inch diameter.

It comprises a steel tubular casing, forming the cylinder,

a piston hammer, inlet, exhaust, and regulating valves, etc., as shown in the illustration. There is a hollow handle through which compressed air is admitted via the trigger valve " A ".

The riveting or hammering action results from the rapid reciprocating motion of the piston hammer, which on its forward travel or stroke imparts a blow to the snap-tool.

The cycle of operation is as follows :—

Assume the piston to be at the end of its forward stroke (as shown in the illustration, Fig. 108)—

(1) Compressed air is admitted by the depression of the trigger valve " A ".

(2) Valve " B " is opened by the pressure of air through the " U " duct (Fig. 108b), and thus closes the passage to the top end of the cylinder, but opens the " transfer " duct to " D " at the cylinder's lower end through which the air passes.

(3) The piston under air pressure is driven to the top end of the cylinder.

(4) During its upward travel the piston passes and uncovers exhaust port " E ", releasing the air to the atmosphere.

(5) Valve " B " closes, and in so doing uncovers duct " L " at the cylinder top, but covers the transfer duct to " D " (Fig. 108a).

(6) The piston is now driven rapidly to the bottom of the cylinder, and strikes a sharp blow on the snap.

(7) During its downward travel the piston passes and uncovers exhaust port " C ", releasing the compressed air to the atmosphere.

(8) The air pressure in the cylinder is now decreased, thus valve duct " L " is again closed by valve " B ", but the transfer duct to " D " is opened so the piston is again driven upwards and the cycle of operation is repeated.

It will be noted valve " B " is of a " floating plate-disc " type.

The machine is stopped merely by releasing the valve " A " trigger, which is then automatically closed by its spring and cuts off the air supply.

When using a pneumatic riveting tool the operator should grip the handle with his right hand, with the thumb conveniently positioned for operating the trigger. His left hand should be used to steady or guide the machine by gripping it around the lower end of its casing.

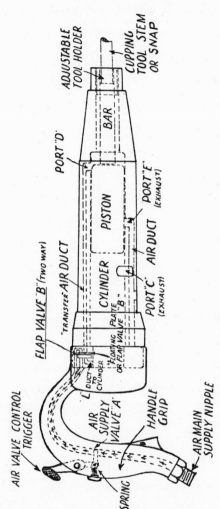

Fig. 108.—Hand-type Pneumatic Riveting Tool.

After the hot rivet has been placed in position in the work and firmly supported, the tool is next applied, the trigger valve opened, and riveting started.

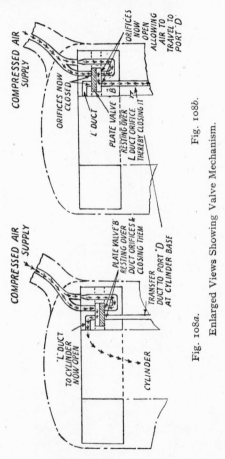

Fig. 108b.

Fig. 108a.

Enlarged Views Showing Valve Mechanism.

It should be firmly pressed over the rivet end, and as formation of the head develops, the operator should move the tool around it in a circular motion. This is to ensure that the snap presses well on to and all around the edge of the rivet head.

The last operation is important in order to cause the edges of the rivet head to bind well and form close contact with the work-piece.

Other patterns of pneumatic riveters are fitted with valves which differ from the " floating plate-disc " illustrated. Some machines employ a small " cartridge "-type spool which oscillates to and fro horizontally across the top of the cylinder, thus regulating the air supply to the top and bottom ends of the cylinder alternately.

Further types have a groove cut around the circumference of the piston, which regulates the admission or exit of air as it passes various " ports " situated at predetermined positions along the cylinder walls.

Riveting tools of the foregoing descriptions are sometimes referred to by operators as " *pom-poms* ".

The person who operates a pneumatic riveting machine is called a " *riveter* ", and the man or boy who supports the rivet during the process of its head formation is known as a " *holder-up* " or a " *bolsterer* ".

CHAPTER IX

SOLDERING TOOLS

SOLDERING IRON (Fig. 109)

ALTHOUGH this tool is called an " iron ", only the shaft spindle is made of iron or more usually of mild steel. The business end of a soldering iron is invariably made of solid copper. The reason for this is that copper retains its heat for considerable periods, and because it can be easily " tinned ". Tinning is the process of giving the copper end a thin coat of solder.

A soldering iron usually has a wooden handle attached to the shaft end, otherwise the heated copper would conduct some of its heat to the metal shaft and it would be difficult to hold by hand.

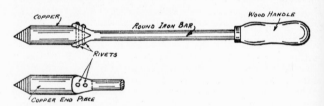

Fig. 109.—Soldering Iron (straight type).

For general soldering work, the larger the volume of copper fixed to the end of the iron, the better ; as it does not need heating up so frequently. For small or light work, however, a relatively small copper end is found more convenient for use, but it requires heating more often.

Soldering is the action of joining together two pieces of metal (one metal to another), or of joining alloys by the help of another more soluble alloy. The soluble alloy used is called " soft solder ". It consists chiefly of a mixture of tin and lead. Solder is usually supplied in rod or bar form.

For all soldering work it is essential that the work

should first of all be thoroughly cleaned of all rust, dirt, grease, oil, etc.   For removing oil or grease, a piece of rag which has been soaked in petrol or paraffin can be used, but all traces of the latter must then be cleaned off with a clean, dry rag.   Rust can be gently filed off, or removed by rubbing with a piece of emery cloth, sand paper, etc.   In any case, it is often an advantage to file the article gently so as to expose the raw metal which is to be soldered.

Assuming the " cleaning-up " process has been effectively completed, the next step is to have close at hand some method of heating the soldering iron.   A coal, coke, or gas fire is quite effective, or a " Bunsen " flame or " blow lamp " can be employed.

The soldering iron, which must also be clean, should be heated sufficiently (but not until it is red hot) until it can be tinned, *i.e.*, when the rod of solder is applied to the copper end it should just stick to or adhere to the copper without running off it.

Note : The correct heat will soon be found by experience.

Some sort of " flux " should also be at hand.   There are several brands of special resinous mixtures supplied for fluxes.   Killed spirits of salts (a saturated solution of zinc in hydrochloric acid) forms one of the most effective, but it is not quite so convenient for storage as the resin-compound mixtures.   The latter are supplied in tins or collapsible tubes.

Spirits of salts, being a strong acid, must be stored in either an earthenware or glass container until it is " killed " by the addition of zinc.   To kill the strong spirits, small bits of zinc are added until " saturation " is obtained. When adding zinc, effervescence or " fizzing " occurs, and small bubbles of gas are given off.   When sufficient zinc has been added to reach saturation point the fizzing stops, and the flux is ready for use.

When tinning the soldering iron, it should have its copper end dipped, whilst hot, into the flux.   The action of using a flux is to prevent " oxidisation " taking place, as this prevents the solder adhering to the work.   Having prepared the soldering iron by heating and tinning it, the work should be given a coat of the flux and the copper tip of the iron should be applied to the spot to be soldered ; it should also be gently pressed along the work in order to heat it.

The rod of solder should next be brought into contact with the copper tip of the soldering iron, which melts the solder and causes it to " run ".

By gently moving the hot iron along the work in order to evenly distribute the solder, the joint is made, and it should then be allowed to cool. Any excess of solder can be filed off when cool.

If the soldering iron will not " take " the solder when it is applied to it, the iron is either not clean, too hot, or too cold. If the iron is too hot the solder " splutters " and " flies off ". If too cold, it will not melt or run.

When using killed spirits of salts as a flux, the user should endeavour to keep his head clear of the fumes which are given off in order to avoid inhaling them. These tend to cause the eyes and nose to run, similar to the effects of a bad head cold. The resin fluxes do not affect the user in this manner.

All traces of flux must be cleaned off on completion of the work. This especially applies to soldering work on any electrical apparatus. Many people prefer to use the spirits as a flux, and find them most effective. Rods of solder can now be obtained, having a central hole in which is inserted a core of flux. This is convenient and obviates the necessity of using a separate flux. Although no doubt these are very handy for small soldering jobs, for larger work they are not quite so effective as the older method of a separate container of either spirits of salts or a resin-basis flux mixture.

In addition to the hand type of soldering iron shown in Fig. 109, electrically heated types are now extensively used. These have an electric element coupled to the iron which heats up the copper end piece, thus obviating the necessity of individual heating from a fire or gas flame, etc. The electric soldering iron therefore saves valuable time in not having to be constantly re-heated.

There are several types of electrically heated irons on the market, most of which are quite good, especially for small soldering jobs. The heat dissipated by the iron on large work, however, often cannot be replaced fast enough by the small type of iron. Of course, the electric soldering iron can be used only where a suitable electricity supply is available.

In addition to the foregoing types, there are others on the market, of which the following are worthy of mention.

GAS-TYPE SOLDERING IRON (Fig. 110)

This has a Bunsen gas jet positioned inside the end-piece, which is made either of copper or cast iron, but in all cases the tip of the end-piece is made of copper. It has a hollow handle stem down which passes the gas-supply pipe. It is coupled by a flexible tube to a gas-supply point.

The end-piece in which the Bunsen burner is fixed has air holes around its perimeter so that air can mix with the gas and allow combustion. The Bunsen flame heats up the copper tip of the soldering iron ready for its application to the work-piece.

The strength of the Bunsen flame can be regulated as required and the supply of gas can be adjusted by the gas

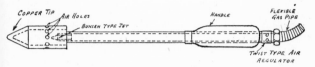

Fig. 110.—Gas-type Soldering Iron.

tap, which forms the main gas supply. Near the handle end is an extra air slide, by means of which further adjustment can be made to the air supply.

This kind of tool is very useful in cases where no electric supply is available. It is perhaps most suitable where many soldering jobs have to be done, rather than having to frequently re-heat an ordinary hand type of iron. It is of rather heavy construction, and may be found a little cumbersome in use until one gets accustomed to it.

Fig. 111.—Cartridge-type Soldering Iron.

CARTRIDGE-TYPE SOLDERING IRON (Fig. 111)

An interesting type of soldering iron has recently been introduced from America. It is known as the " Quick-shot " iron. The chief characteristic which is claimed for this tool is that it requires no external heating whatever. It is independent of any gas or electric supply, or any flame type of heat generator.

The principle of the Quick-shot iron's operation is that the heat is supplied by a cartridge composed of some magnesium compound mixture. At the soldering end of the iron, immediately next to the copper end-tip, is a casing into which a cartridge is fixed.

Through the handle and down the stem, a bar protrudes, which has a " striker " on its lower end. This striker engages with the cartridge and, when it is desired to heat the tool, the plunger, which is positioned at the handle end, is pushed inwards and this action ignites the magnesium cartridge. The latter rapidly generates heat, which heats up the copper end-tip in five seconds. It is also claimed that sufficient heat is generated, and that the temperature is maintained for a period of seven minutes, which is sufficient time to complete the average soldering job.

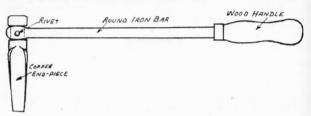

Fig. 112.—Angular Soldering Iron.

Several different sizes of copper end-tips are supplied, each of which can be screwed into the chamber of the cartridge-holder. This procedure thus provides for soldering both small and large work-pieces by merely fitting the most conveniently sized copper end-piece.

A supply of cartridges can be stocked, which allows a soldering repair job to be carried out at a few minutes notice, instead of having to wait for an electric or gas-fired iron to heat up.

This type of tool should be found of great value to motorists, who, in case of breakdown, can do any soldering job by the roadside quite easily and in a few minutes.

It should also be found very handy in cases where soldering work has to be carried out during the installation of plant or machinery on new sites, where no electric power is available at that stage of the job, and where a blow-lamp or fire of some kind would otherwise have to be used to heat an ordinary hand type of iron.

ANGULAR SOLDERING IRON (Fig. 112)

This is an ordinary hand type of tool, and in general construction it is very similar to that shown in Fig. 109, except that the copper end-piece is set at right angles to the handle shank.   For some classes of work, it is found more convenient in use, as the pointed end can be applied to the work without unduly straining the holder's hand or arm, as would be the case with a straight type of iron.   The angular soldering iron is often called a " hatchet " type.

It is used in exactly the same way as any other hand type of soldering iron.   After being accustomed to using the straight type of soldering iron, the " angular " pattern may be found a little awkward to use, but after a little practice the user will soon get acquainted with it.   It will be found very useful for penetrating inside the otherwise awkward corners of jobs of a hollow nature.

CHAPTER X

HAND WELDING AND CUTTING TOOLS

PROBABLY the most elementary form of welding tools are those used by the blacksmith when welding together two pieces of iron or steel. In doing do, the blacksmith uses his hand hammer extensively in order to beat together the two pieces of metal. Plain welding has been introduced to the reader in Chapter VII under the heading of " Forging Tools ".

In order to effect a weld, the blacksmith first of all heats the two pieces of metal up to their fusing temperature. He then places them together whilst they are hot and " beats " them until they are fused into one piece. During the beating action, small scales form on the outer faces of the work-piece. These scales are due to some of the carbon being expelled from the hot metal. Perhaps they are more noticeable as the work-piece " cools off " slightly from its red heat. The action of expelling some of the carbon assists in the two pieces of metal being joined together. It also tends to make the material more " pliable " or " ductile ", and in consequence reduces its brittleness. Small forgings are often welded together in this manner, particularly in the formation and welding of hand-forged chain links.

Of recent years, however, other types of welding have rapidly advanced and been perfected. These later methods of welding can be applied to a far wider field than those which are employed by the blacksmith by the use of his hand-hammer.

Some of these other types of welding will now be introduced to the reader, in order to acquaint him with a few of the elementary tools concerned.

OXY-ACETYLENE WELDING BLOW-PIPE (Fig. 113)

If suitable proportions of oxygen and acetylene gases are mixed together and fed through the fine nozzle of a blow-pipe, the combined gas, when ignited, rapidly generates intense heat.

The heat can quickly bring metals up to their fusing temperatures when they can then be successfully joined—or welded—together.

The oxygen and acetylene gases used for welding work are supplied in highly compressed form.   They are contained in large cylinders, which are fitted with pressure indicators, outlet pressure gauges, and various supply valves.

The oxygen cylinder is painted black, and the acetylene cylinder is painted maroon (or brick red).   The cylinders are painted in these colours so as to be easily recognised, and thus avoid confusion or accidents.

The gases are fed by flexible rubber tubes of suitable length to a blow-pipe, which is the tool used for the regulating of the flame used to carry out the welding work.   The blow-pipe which is shown in the illustration of Fig. 113 has

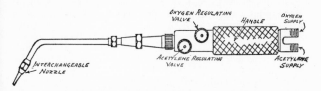

Fig. 113.—Oxy-acetylene Welding Blow-pipe.

two valves positioned adjacent to the handle.   By manipulating these valves, a suitable combination of the two gases can be effected.

At the extreme end of the blow-pipe a nozzle is fitted, which controls the jet flame.   Various sizes of nozzles are supplied for the different sizes of welding work involved. These are easily interchanged by merely unscrewing and replacing by one of a different size.

For welding wrought iron or mild steel no flux is required in most cases, but for welding other metals or alloys some sort of flux should be used.   A suitable flux is one which is composed of borax powder, fine powdered silica sand and iron filings.

In order to form a welded joint between two pieces of wrought iron or mild steel their ends or faces which are to be united must first of all be cleaned so as to exclude any scale or rust.   A wire brush or a file can be used for this purpose.

Various forms of joints are used, some of which are of a
" V " notch type or of a taper formation, etc.

The two pieces are placed in position, and the jet flame
brought to bear on the joint. A piece of special welding
wire is then placed in the flame and the heat melts it
(similarly to when doing soldering work). The "drop-
pings" from the melted welding wire are deposited on the
joint and assist in forming the weld.

When any oxy-acetylene welding is being carried out, the
operator (or " welder ", as he is called) must use goggles
with dark-blue or green-coloured lenses to protect his eyesight.

Some types of blow-pipes are fitted with water-cooled
heads. Such pipes have a water-jacket which is supplied
with a circulation of cold water in order to keep them
relatively cool. These water-cooled blow-pipes are required
only for special jobs requiring long and continuous heating.

Blow-pipes, when in use, are sometimes liable to " back-
fire ", or the flame is liable to " blow back " into the supply
pipe. When this occurs, the operator should, first of all,
shut off the oxygen valve and then close the acetylene valve.
After waiting a few moments for the gases to clear from the
pipes, he can turn on the valves again and re-ignite the gas
at the nozzle jet end.

After a weld has been made, it should be allowed to cool
off slowly, and should never be dipped in water to " quench "
it quickly.

OXY-ACETYLENE CUTTING-TOOL OR BURNER
(Fig. 114)

In order to cut metals by burning through them, a special
type of blow-pipe is used. This is called a cutting blow-
pipe or " burner ". It is coupled up to similar oxygen and
acetylene cylinders to those previously described.

A cutting-tool—or burner—differs from a welding blow-
pipe chiefly by reason of it having two separate gas supplies
at the nozzle end. One supply forms the heating flame, and
the actual cutting flame is supplied by a jet of oxygen.
By the use of such a blow-pipe, metal plates or sheets can
be cut rapidly by burning a thin line through them.

To use a cutting blow-pipe, the oxygen and acetylene
gases are suitably mixed by regulating the supply valves.
The mixture is then fed to an outer ring of the nozzle and
forms the " pre-heating " flame. Through the centre of
this outer ring, a fine jet of pure oxygen is supplied from a

by-pass pipe in the tool. The central jet flame produces terrific heat with which the actual cutting of the plate is done. The cutting flame must be retained at a certain even distance from the metal and as the cutting or " burning " proceeds the tool is moved along longitudinally.

Some types of blow-pipe nozzles are fitted with two small rollers—one each side of the nozzle to ensure the keeping of an even distance from the work. The nozzle should be held at right angles to the work.

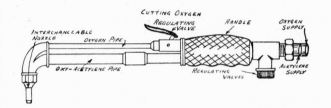

Fig. 114.—Oxy-acetylene Cutting-tool or Burner.

ELECTRIC-WELDING ELECTRODE HOLDER (Fig. 115)

Welding work can be effected by using an electric current. If such a current is applied to a work-piece and an electrode which is coupled up to the same supply is brought close to the work-piece, a flame or " arc " is caused.

A heavy current is used and the " electrode " is formed of welding wire (usually made of soft iron) which can be easily melted or fused.

When the arc is formed, the heat generated causes the electrode to melt, and particles of the latter are deposited on to the joint which is also heated up to its fusing point by the heat from the arc flame.

The electrode rods are specially made for the purpose, and take the form of thin, round bars with a special coating on their outer faces. This special coating greatly assists in forming the weld joint.

In order to hold the electrode a special tool must be used called an " electrode-holder ", and a typical one is shown in Fig. 115. From the illustration it will be seen that the tool somewhat resembles a pair of flat-nosed pliers with a spring between the handles and with one handle projecting.

The latter is covered with thick insulating material, for it is by this handle that the operator holds the tool. The smaller handle, which is also insulated, is used to release the grip of the jaws, when inserting a fresh electrode.

The electric-supply cable which carries the current passes through the core of the larger insulated handle, and along the arm to the copper-faced jaws, and thus conveys the electric current to the electrode which is clamped between the jaws.

When carrying out welding by the electric process, special protective clothing should be worn, consisting essentially of a leather, or fibre, head shield to which is fitted a coloured glass visor, in order to protect the eyesight of the user. Special heavy leather gloves and a stout leather apron should also be worn.

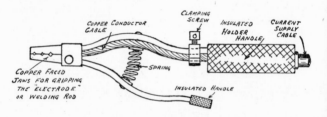

Fig. 115.—Electric-welding Electrode Holder.

Note : The principles of both oxy-acetylene and electric welding are dealt with more fully in Volume III, but the foregoing paragraphs and illustrations are intended solely to introduce to the reader some of the elementary tools to be encountered in such work.

CHAPTER XI

MISCELLANEOUS TOOLS

WET GRINDSTONE (Fig. 116)

THIS is a tool used for grinding or sharpening, and consists of a wheel which is hewn out of solid natural stone. Usually

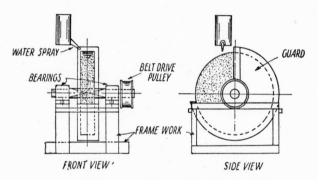

Fig. 116.—Wet Grindstone.

it has a square hole in the centre, in which is mounted a spindle on which the wheel rotates. The spindle runs in bearings, and is either turned round slowly by hand or by a belt and pulley. Immediately above the rotating stone is a water-sprinkler, by means of which water can be sprayed on to, and allowed to trickle down, the perimeter of the stone. This water forms a lubricant and prevents the article which is being ground from getting hot and consequently from having its edge softened.

Although the grindstone has been superseded by "emery" types of grinding wheels in many workshops, it is still favoured by some workmen for the sharpening of chisels and small tools.

EMERY WHEEL GRINDER (Fig. 117)

This is another type of grindstone, but, instead of being made of natural stone, it is made of a mixture of carbide, coke dust, and other materials which are " fused " or " cemented " together at a very high temperature. Whilst in a hot, molten state it can be cast or moulded into any desired shape. Wheels of this material are made in several degrees of " coarseness ", and are used for various classes of grinding work. These wheels rotate at very much higher speeds than the grindstone previously mentioned. They are used " dry " for some work and sprayed with water for others.

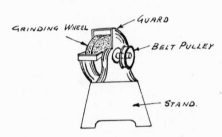

Fig. 117.—Emery Wheel Grinder.

A cold chisel (and similar tools) should never be sharpened on a dry emery wheel, as its speed of rotation rapidly creates friction, which in turn generates heat. The heat draws out the temper from the steel and softens the point.

When using an emery wheel " sparks " fly off in all directions, and as a precaution it is usually fitted with a guard which covers all but the actual part of the wheel where the work is placed to be ground.

Emery wheels of various grades are extensively used in machine shops for finishing off work. Emery wheels are used for cleaning up rough castings in a foundry, and in constructional shops, etc., for finishing-off rough-sawn sections of steel joists, or for quickly grinding down a piece of steel plate, etc.

ENGINEER'S OIL-CAN (Fig. 118)

Although possibly a simple tool, or appliance, an oil-can is very important. Much wear and tear can be saved by the regular application of an oil-can to any machine, especially if the latter is one which consists of several moving parts, such as shafts, wheels, or reciprocating levers, etc.

An engineer's oil-can consists of a container which is fitted with a removable filler cap, as shown in the illustration. A handle of " loop " form is provided as shown.   Near the handle, at the top of the container, is fitted a " press-button " type of valve which regulates the flow of oil from the container to the spout.

Some oil-cans have fixed spouts, while others have spouts of a detachable type, similar to that which is illustrated. In this case the spout is retained in position by a screwed " gland " or " nipple ", by means of which the spout can be readily detached for cleaning purposes.

Oil-cans vary in size from those which will hold half a

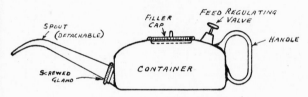

Fig. 118.—Engineer's-type Oil-can.

pint of oil up to those capable of holding two pints.   They are usually made of tinned sheet steel, but may also be obtained in copper or sheet brass.

PUMP-TYPE GREASE-GUN (Fig. 119)

A grease-gun is used to " force " grease under pressure into the " nipples " of machinery.   The nipples are the appliances by means of which the grease is supplied to the various parts of the machine concerned.   They are made to various standard sizes, and are screwed into the machine at the various positions which require lubricating from time to time.

A grease-nipple usually contains a small ball-type valve, which although it will allow grease to enter will not allow it to escape or work outwards.   Such a valve is called a non-return type.

The grease-gun consists of a cylindrical container (or

" barrel "), as is shown in Fig. 119. The barrel is provided with screwed cover-caps at both of its ends. At the rear end of the barrel, a plunger rod projects. This rod has a handle at its extreme end. The opposite end of the plunger rod is fitted with a leather plunger and a washer, which are very similar to those of an ordinary cycle-pump. At the front end of the barrel, a tapered guide sleeve is fitted.

From the sleeve a feed-pipe is arranged, as shown in the illustration. At the delivery end of the feed-pipe an adaptor is arranged, which is suitable for fitting over the entrance of the nipple concerned.

In order to use a grease-gun, it must, of course, first of all be filled with grease in the manner indicated.

The filling of the container or barrel is effected by un-screwing and removing the front cover-cap, together with the feed-pipe assembly. The plunger should then be

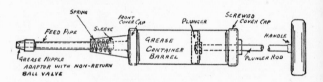

Fig. 119.—Pump-type Grease-gun.

positioned at the extreme handle end of the barrel. Grease is then applied to the open end of the barrel until the latter is filled. The front cover-cap, complete with its feed-pipe assembly, is then replaced; the cap is screwed up, and the grease-gun is fully " charged " ready for use.

The adaptor end is then placed in position over the nipple concerned, and the barrel should be held by one hand of the operator. Pressure should then be applied to the grease-gun handle by the operator's other hand, which compresses the grease inside the barrel, and forces it down the feed-pipe to the nipple.

While still retaining pressure on the handle, the grease-gun barrel (to which the sleeve is attached) should be pushed bodily, or forced along the feed-pipe in the direction of the nipple position. This action forces or " pumps " the grease through the nipple ball-valve, from which it cannot return. The grease is thus fed to the part of the machine requiring lubrication.

During the action of forcing the barrel sleeve along the feed-pipe, the spring will have been fully extended, owing to its being attached to the end of the pipe. At the end of the downward " pump-stroke " the spring " reacts ", and returns the barrel to its previous position, as the pressure is also released by the operator's hands after the completion of the stroke. Hand pressure is again applied, and the downward pump-stroke repeated. This forces more grease through the nipple and these actions should be repeated until sufficient grease has been applied.

After each successive stroke the plunger will have been forced farther down the container barrel, as more and more grease will have been discharged. When the plunger has reached its full extent down the barrel, the contents will have been emptied, and the barrel will require replenishing with grease, as previously described. This stage can be ascertained by noticing when the handle of the plunger rod reaches a position near the rear cover-cap of the barrel.

Some types of grease-gun, instead of having straight adaptors, have adaptors of an " inclined " nature, in order to fit angular-type nipples.

Other types of grease-gun may have a flexible steel pipe positioned between their sleeves and the nipple adaptors. This provision is made so that they can be used to service nipples which are positioned in parts of machinery found difficult of access with straight, fixed steel feed pipes.

Another type of grease-gun has a screwed pattern plunger-rod instead of a plain rod. Pressure is applied to this type by rotating the handle of the screwed rod, which causes the rod to screw into the cap and the plunger slowly to descend the barrel. No pressure need then be exerted by the hand on the plunger-rod handle while the barrel sleeve is being forced down the feed-pipe. The screwed type of plunger-rod grease-gun is somewhat slower in operation than that which is illustrated in Fig. 119.

INDEX

Nannie Barron.
Christmas 1900.

SCOTTISH NATIONAL DANCES:

A PRACTICAL HANDBOOK.

WITH ILLUSTRATIVE DIAGRAMS.

BY

J. GRAHAMSLEY ATKINSON, Jun.

(Late Manager for Mr and Miss Atkinson, Teachers
of Dancing to the Royal Family).

117 GEORGE STREET, EDINBURGH.
1900.

PRINTERS
WESTWOOD
AND SON.
CUPAR-FIFE

CONTENTS.

PREFACE.

* * *

"Among the splendid array of works of living and dead authors whose talents have been devoted to an illustration of the history, the topography, the statistics, and other peculiarities of Scotland, it is somewhat surprising that no one is found wholly or even partially dedicated to the exposition of its institutions."

Chambers' "Book of Scotland."

The above quotation forms an apposite prelude to these Letters.

Scottish traditions, customs, institutions, history, law, literature, art, songs, and dance music, have been investigated by experts, and the results of their investigations preserved in book form for the information of inquirers and for historical comparison, but Scottish National Dances have, so far, received little attention.

From an historical point of view this is much to be regretted. Scottish National Dances have, no doubt, lost their original construction and meaning in obedience to that law which operates when an art is left for generations without record or authoritative standards.

From a practical point of view, this absence of record may be said to have benefitted the art, since, instead of dances stereotyped alike in form and detail, we possess dances fundamentally the same, yet exhibiting structural differences due to their independent and contemporaneous development in different localities—localities not only separated geographically, but by limited intercourse, and especially by that clannishness which is a characteristic of the Scottish people. Each clan guarded its customs, songs, music, and dances as jealously as its patronymic or its honour.

It is true that travelling facilities, more frequent intercourse, and an ubiquitous press have done much to remove the outward and visible signs of this passion for long established local custom, but it requires only the words of the familiar song, the sound of the familiar tune, or the sight of the familiar dance to brighten the eye, quicken the pulse, transform the countenance, and rekindle the latent fire that lies hid in the breast of a Scotsman.

All admirers of Scottish National Dances must regret there is no work to which the inquirer may turn for information regarding their detail and construction. There has been handed down to us by tradition a number of

dances possessing qualities of a high order—
whether considered from an artistic point of
view, or as embodying the character of a
people, or as physical exercise and training.

Some writers have devoted a few sentences
to this subject, but by none are the dances
reduced to their elements and again recombined
to show their construction.

There is one modern writer (Mrs Lilly
Grove, F.R.G.S., The Badmington Library—
Dancing) who explains the absence of authori-
ties, as follows (p. 180):—

"The reason of this poverty of description is that the
Scot, whilst practising the musical arts, had not yet (pre-
Reformation times) reached such a height of civilization
as to pen treatises on any of the arts, dancing among
them."

Then Mrs Lilly Grove, F.R.G.S., goes on to
show the enormous strides civilization has made
since then by handing down to posterity the
following choice specimen of descriptive
English (p. 188):—

"One may suitably begin the more particular
account of Scottish national dances by the description
of the step peculiar to the Highlands, and known as the
'Highland Fling.' . . . The term 'fling' expresses
the kick which characterises the step.

"When a horse kicks by merely raising one leg and
striking with it, he is said 'in groom's parlance' to 'fling
like a cow.' This is what the Highland dancer does ; he

dances on each leg alternately, and *flings* the other one
in front and behind."

How worthy of preservation is this allusion
to the stable and the byre. How eminently
satisfactory as description is the above quotation
in so prominent a work—one of a series the
writers of which, the preface tells us, "are
thoroughly masters of the subjects of which
they treat." Surely the inexperienced reader
cannot fail to carry away a clear and an exalted
idea of "the step peculiar to the Highlands."
Again (p. 189), Mrs Lilly Grove, F.R.G.S.,
shows what a keen observer she has been
during her five years' (p. 6) study of dancing—
thus :—

> "The Scot arrives at the dance floor as he would
> the drill square, and he dances till he is tired out, rarely
> looking at his lady partner (if he has one), and, in fact,
> caring not with whom he dances."

The discriminating reader will at once perceive
Mrs Lilly Grove, F.R.G.S., must have made a
slip of the pen when declaring she studied the
subject for five years. It is manifest she must
have meant five minutes, but even at this
estimate, the reflection remains that she utilised
the time to little purpose.

This experienced authority now goes on to
what is intended to be description of detail.
She begins by describing the figure of the Reel.

Here is the description by Mrs Lilly Grove,
F.R.G.S. :—

> "The figure of the Reel is perhaps the most beautiful
> that can be exhibited. Hogarth exemplifies it as the line
> of beauty."

That is all. How lucid. How skilfully calcu-
lated to raise in an inquirer's mind the exact
detail and construction of the figure of the Reel.
This is all the description of detail Mrs Lilly
Grove, F.R.G.S., vouchsafes as the result of her
five years' study of dancing. Feeling, however,
she must say more, and being presumably out
of her depth, she clutches at a Mr Francis
Peacock, who wrote a few sketches on dancing
about the beginning of this century. She calmly
offers her readers a "brief description" (p. 192)
of Mr Peacock's steps, but the flavour of the
stable and the influence of the "cow-like kicks"
still remain, therefore she succeeds only in
rendering Mr Peacock's rather obscure descriptions absolutely unrecognisable. To crown all,
she gravely writes :—

> "If the reader will carefully study the above
> descriptions (namely, her version of Mr Peacock's steps),
> he will in time acquire some notion of the fascinating
> Scottish Reel, and will be better able to understand the
> evolutions should he have the good fortune to see it well
> performed."

The foregoing sentence, following as it does

Mrs Lilly Grove's burlesque, would admirably suit a screaming farce, but it is nothing less than an insult to the reader when seriously written in a work, which is one of a series, claiming to be written by experts.

The following Letters have been written in the hope that one small step in the right direction has been made. They are written by one who has been long engaged in the teaching of Scottish Dancing, and who is a true admirer thereof.

The endeavour throughout has been to preserve the form and construction of the dances as they obtain now.

Someone wielding an able pen may yet undertake the task of compiling an historical as well as a practical work on the larger question of Scottish National Dancing. The following Letters are restricted to the dancing of Scottish Reels.

All who know Scottish National Dances will admit they bear the impress, or are the exponents of a manly, vigorous, and practical people—sound and healthy both in body and in mind. There is neither voluptuousness nor langour about them. They typify activity, alertness, presence of mind, fertility of resource, independence, attention to minutiæ, the concession of rights and privileges to others while maintaining those of the individual. These qualities and others are embodied in Scottish Dances. In no class of Scottish Dances are those qualities more strongly manifested than in Scottish Reels. No claim is made to literary merit or to completeness of treatment in these Letters. Indeed, to one who already knows Scottish Dancing, these Letters will appear feeble and insufficient, but enough has been said to give an inquirer into the subject a reasonable conception of the essentials of Reel Dancing. To attempt a fuller description would, in the absence of technical terms, appal the ordinary reader, and to introduce a mass of technical terms, with their definitions, would convert this little work into a "Treatise on Dancing," which it does not claim to be.

These Letters treat, in plain every-day language, of one small department of Scottish Dancing. When a confession of this kind is made it is customary to inform the reader where a fuller if not complete account of the subject may be found. In the present case it is to be regretted there is no complete account available.

The epistolary form has been chosen as the nearest approach to a personal method of instruction.

The letters are addressed to a gentleman, but where there are any differences in the duties of the lady and of the gentleman, these are explained as they occur.

The author is greatly indebted to Mr Edward Scott, of London, for many valuable suggestions which have been embodied in this little work.

117 GEORGE STREET,
    EDINBURGH, 1900.

LETTER I.

❋ ❋ ❋

My Dear A——

      In beginning this series of Letters on the dancing of Scottish Reels, I am desirous you should have a clear idea of the limits of the inquiry upon which we are about to enter. Some think the term Reel has a specific meaning, and is used to indicate a dance of definite form and construction. This, however, is not the case. The term Reel has a wider meaning when used in dancing. It embraces several dances which though having a strong family likeness yet differ in construction. The term Reel is used as a common term, and includes various reels which are further defined by specific titles, such as :—

    The Reel and Strathspey or "Foursome,"
    The Reel of Tulloch,
    The Reel of Three,
    The Reel of Eight or "Eightsome,"

and others less generally known. Each of

these Reels, again, has some variation in form that is generally recognised—there are unrecognised variations as well. It would be unprofitable to attempt any description of unrecognised forms—I will therefore treat only of those which have been handed down as traditionally correct, and which are now danced by those who know what Scottish dancing is.

The purpose of these Letters being practical, I will not attempt an historical account of the Reel—to do so would plunge us into a sea of doubt and conjecture. Whether the Reel be indigenous to Britain or introduced from without? whether it is as ancient as tradition or as recent as the Scandinavian invasion of these Isles?— and many other questions—are questions that may interest the historian and antiquary, but decisions one way or the other would not aid you in acquiring the dance of to-day.

Whatever be the origin or age of these dances, they are now the Scottish National Dances, and are woven into the nature of the people.

In my attempt to place before you the construction of the various forms of the Reel, I do not intend to give a picture in which every rich detail is reproduced—this would entail a mass of matter and a series of digressions, tending to obscure rather than to define. I will adopt a more effective method. I will divest the dance I wish to describe of all but the barest essentials, and, using these, will give the *form* and *construction* or *framework* of the dance.

Once you clearly comprehend the *form*, you will perceive something more is necessary than the barest essentials to give interest and variety. It is true a skilled dancer may and does use many beautiful steps, but he does not in the least alter the fundamental form of the dance—he embellishes.

In describing the dances I will, where possible, use ordinary language, but as the knowledge of a few simple technical terms will enable me to present the dances in broad, clear outlines, I hope you will master them as they occur.

Let me revive your recollection of the five fundamental positions in dancing :—

1st. Bring the heels close together—the feet being conveniently turned out.

2nd. From the first position extend one foot directly sideward, so that the toe rests (without weight) on the floor at the full extension of the limb.

3rd. One foot is half-crossed before or behind the other—the feet being conveniently turned out.

4th. Extend one foot directly before or behind the other, so that the toe rests (without weight) on the floor at the full extension of the limb.

5th. One foot is *completely* crossed before or behind the other—the feet being conveniently turned out.

NOTE.—The 1st, 3rd, and 5th positions are called *close* positions, the 2nd and 4th *open*.

In the open positions I have said "extend one foot in a given direction with the toe resting lightly on the floor"—the other foot, by inference, completely on the floor. Suppose, while making an open position, you rise to the toe of the supporting foot, you still form 2nd or 4th as the case may be. Further, suppose the toe of the extended foot is raised an inch or two from the floor, you still form the 2nd or 4th positions. Whenever I have occasion to require the extended foot raised from the floor in an open position, I will say 2nd or 4th position *raised*. You must be careful that the foot is raised by an action of the whole limb, and not by raising the fore part of the foot only. In close positions, also, you may rise to the toes—in doing so the positions are still 1st, 3rd, or 5th as the case may be. You may take it for granted throughout these Letters that you *always* form the positions, both open and close, with a powerful upward pressure from the toes.

I have treated you mercifully with regard to the number of positions—beside me lies a work in which the writer not only gives five fundamental positions, but in addition he gives *twenty-one* derivatives. Upon this broad foundation he builds an elaborate structure—well suited, no doubt, to the requirements of a professional reader, but bewildering in its complexity to the non-professional.

Do not despise these simple positions—all the movements I have to describe begin from and end in one or other of them.

The scheme I intend to follow out in these Letters is as follows :—Letter II. will be devoted to Reel and Strathspey music; then I will devote one Letter to each well-recognised form of Reel, another to less-known Reels, and conclude with notes, general observations, and a few characteristic steps for ladies and gentlemen.

In describing the various Reels, I have found it convenient to make each Reel a stepping stone to the next. It will, therefore, be necessary to acquire the dances in their order. To have done otherwise would have necessitated the repetition in each Letter of parts of those preceding.

FAREWELL.

LETTER II.

* * *

My Dear A——

There is much confusion in the minds of many with regard to the correct *Tempo* (that is, rate of playing) and *Accent* of Reels and Strathspeys, and as they differ most markedly in these respects, I invite your close attention to this Letter.

Reels and Strathspeys differ so much in tempo and accent that a Scot can hardly realise anyone failing to recognise their characteristic differences, but as there are books of Reel and Strathspey tunes in which the names Reel and Strathspey are wrongly applied, I cannot wonder at the doubts and mistakes of an inquirer when information is sought therefrom. Not only are the names in some cases misapplied, but the tunes are sometimes falsely written, and there is seldom any sign to indicate how fast or how slow they should be played, or how they should be accented to suit the purpose of the associated dances. I will endeavour to aid you here.

For the purpose of these Letters we have no class of Scottish music to consider except Reels and Strathspeys.

Reels and Strathspeys have some points in common.

1st. They are for the greater part in common time.

2nd. They are composed of two eight-bar parts or their equivalent—that is, there may be one four-bar part repeated *plus* one eight-bar part, or two four-bar parts each repeated. For the purpose of the dances there must be two parts of eight bars each—be these parts constructed as they may.

Reels and Strathspeys differ as follows for dancing purposes.

1st. In the rate at which they are played.

(a) In Reels—The Minim=126 Maelzel's Metronome.

(b) In Strathspeys—The Crotchet=176 Maelzel's Metronome.

If you have not a Metronome you can easily make a fairly reliable substitute. Tie a bullet or small weight to a piece of fine twine, and, measuring off nine inches from the weight, suspend it from a fixed point—set the weight in motion. *Two* beats or swings of the weight will occupy *one* bar of a Reel—that is, one minim (not a crotchet) or its equivalent in crotchets, quavers, or semiquavers, to each beat or swing. Measuring off four and a half inches—again suspend the weight and set it in motion. Now *four* beats or swings of the weight will occupy *one* bar of a Strathspey— that is, one crotchet or its equivalent to each beat or swings.

This question of tempo is of much importance. A tune may bear a slight alteration in its tempo without injury, but, if the alteration is considerable, the character of the tune will be destroyed.

2nd. They differ in their accentuation.

(a) In Reels the first and third crotchets in each bar receive a slight emphasis.
You will count each bar thus :—" *One*, and, *two*, and."

(b) In Strathspeys the accent is irregular; each Strathspey seems to have a characteristic accentuation of its own.

For the purpose of dancing I may say that the first and fourth crotchets in each bar receive a slight emphasis.

You will count each bar thus :—" *One*, two, three, *four*."

3rd. They differ in the character of the tunes.

(a) In Reels there is a succession of equal crotchets, or equivalent quavers or semiquavers, giving a running smoothness to the flow of the music and associated dance.

(b) In Strathspeys there are dotted notes and semiquavers—the latter sometimes coming before, sometimes coming after the dotted note. This gives what I may call a "staccato" or "lilting" character to the music, and, of course, to the dance.

The following examples, "Lady Baird" and "Speed the Plough," show the foregoing differences very clearly :—

EXAMPLE OF A STRATHSPEY.

"LADY BAIRD."

Crotchet = 176 M.M.

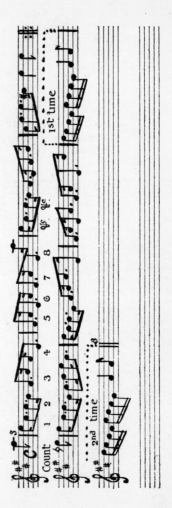

EXAMPLE OF A REEL.

"Speed the Plough."

Minim = 126 M.M.

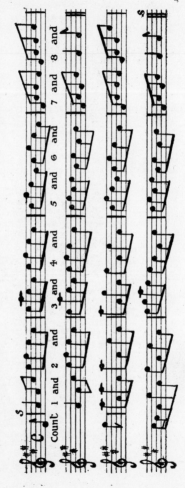

Notice the broken notes throughout the Strathspey, suggestive of a "lilting" in the dance; and again observe the uniformity in the value of the notes in the Reel, suggestive of that running smoothness to which reference has been made. You will further remark—in the tempo formula of the Strathspey—I have used the *crotchet* as the term upon one side of the equation, but in that of the Reel I have used the *minim*. I have used the minim in the latter case because, though there are four crotchets in each bar, yet, in dancing, you have only *two* accented movements in each bar—the minim will, therefore, coincide with the accentuation of the movements made.

In the Strathspey, on the other hand, you have, in dancing, *four* accented movements in each bar—the crotchet, therefore, will coincide with them.

I hope what I have written, and the characteristic examples which have been given, may enable you to recognise Reels and Strathspeys when played.

FAREWELL.

LETTER III.

* * *

My Dear A——

       In Letter I. I gave you the specific titles of these Reels which receive general recognition. I now wish to treat of the Strathspey and Reel in particular; to explain its component parts; to show how the performers place themselves; and to describe the structure of the dance as a whole. It is necessary, in the first place, to define the terms we are about to use; this is all the more necessary since there exists considerable looseness in their use. The Reel and Strathspey—or, to be strictly correct, the Strathspey and Reel—is a composite dance of two distinct parts: first the Strathspey, second the Reel: each part having a rhythm essentially its own.

Why the term "Reel" was chosen to identify a dance in which the Strathspey not only comes first, but has the most distinctive rhythm, I cannot say. The choice has been unfortunate, and escape from its consequences

has been sought in the coining of another
term "the Foursome," which, though not very
apposite, is at least free from the objection of
giving undue prominence to one half of the
dance. Let us then adopt the term "Foursome" to raise in our minds this composite
dance :—

$$\text{Foursome} \quad = \quad \begin{cases} \text{Strathspey.} \\ \text{Reel.} \end{cases}$$

Let us now suppose you are desirous of taking
part in the Foursome. What are the essential
movements or steps you must acquire? You
will no doubt be surprised to learn that if you
become familiar with *four* steps or movements—
two for each division of the dance—you can
take part therein without discredit to yourself,
and without inconvenience to the other three
performers. You must not, however, imagine
that these four steps or movements are all you
require to be a good performer. What I mean
is that you could, if possessed of them, take
part, but without them or substitutes, you could
not.

What are these four movements? They
are :—

1st. Kemshoole (Ceum siubhail)	⎫ For Strathspey	⎫
2nd. A Strathspey step	⎬	⎬ Foursome
3rd. The Chassé	⎫ For Reel	⎭
4th. A Reel step	⎭	

I will describe these movements in their order.

1st. Kemshoole.

Stand on your left foot, the right being held over the left instep.

a {	Slip your right foot forward to 4th position	} one
b {	Draw your left foot up behind the right in 5th position	} two
c {	Again slip right foot forward to 4th position, and pass on to it	} three
d {	Hop upon the right foot, at the same time passing left foot to front over instep of the right	} four

} 1 bar

With the left foot now leading, you repeat a, b, c, d, and so on, with each foot alternately leading, during eight bars. For practice, the Kemshoole should be performed over a curved path, thus :—

Fig. 1

Starting from A (Fig. 1) you would reach the point B at the end of the 4th bar, and regain the point A at the 8th bar.

2nd. A Strathspey step. Let us take a well-known step, which for purposes of reference we will call the 1st Strathspey step :—

Stand on the left foot, the right being placed close behind the left ankle.

a { Hop on the left foot, and extend the right to 2nd position } one

b { Hop on the left foot, and place the right foot close behind calf of left leg } two

c { Hop on the left foot, and place the right foot close in front of the left leg (a little below the knee) } three

d { Again hop on the left foot, and place the right foot behind as at b. } four

} 1 bar

These four actions are repeated with the left foot leading ... ... 1 bar

Again repeat them with the right foot leading 1 bar

And yet again repeat them with the left foot leading, but this time you make one revolution (to the right) upon your own axis while executing the actions ... .. 1 bar

The whole is repeated during the second 4 bars, but you begin with the left foot leading, and turn to the left upon the left foot when finishing ... ... 4 bars

3rd. The Chassé.

Stand on your left foot, the right being placed over the left instep.

a { Slip the right foot forward to 4th position } one

b { Draw the left foot up behind the right in 5th position } and

c { Again slip the right foot forward to 4th position and pass on to it } two

} 1 bar

Passing the left foot to the front you repeat a, b, c, with the left foot leading, and so on during 8 bars.

For practice this would be performed through a curved path as in learning the Kemshoole.

You will here, as in the Kemshoole, reach the point B (Fig. 1) in 4 bars, and regain the point A at the 8th bar.

You must note that you have 32 beats or counts in describing the whole path with the Kemshoole, but with the Chassé movement you describe the same path in 16 beats. It is therefore manifest that if the *space covered* in each case be the same, the *rate of motion* with the Chassé movement will be much quicker than with the Kemshoole. This is intensified by the fact that the Reel is played more quickly than the Strathspey. It is this change of rate of motion, together with the change of rhythm, which give these two movements their distinctive characters.

4th. A Reel step. Let us again take a well-known step, which for purposes of reference we will call the 1st Reel step :—

Stand with the right foot forward in 4th position raised.

a { Bring the right foot smartly down in place of the left, which you immediately extend to 4th position behind } one

b { Bring the left foot smartly in place of the right, which you immediately extend to 4th position in front } two

c { Again bring the right foot down as at "a" } three

d { Hop upon the right foot, and at the same time pass the left foot to 4th position in front raised } four

2 bars

Repeat a, b, c, d, four times, each foot
alternately leading, during 8 bars.

Having now made yourself perfectly familiar
with the above four movements or steps (Kemshoole, 1st Strathspey step, Chassé, and 1st
Reel step), and able, I hope, to perform them
neatly and accurately the moment each is
named, I will now proceed to cast this material
into the form of the dance.

I must now assume you have three friends
(two ladies and one gentleman) able and willing
to aid you in working out the figure of the dance.
This being granted, you will place yourselves
thus :—

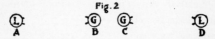

Fig. 2

Two gentlemen back to back, each facing his
partner.

Before describing the structure of the dance,
just a word as to the figure described by the
dancers. During the first eight bars, and every
alternate eight bars, the four dancers describe a
figure which is usually called the figure-eight.
The figure here described is not an exact figure-
eight. I much regret no more suitable term
has been chosen for the figure described in the
Foursome, because a perfect figure-eight is
required in a dance called the Reel of Three.

I am, however, powerless to alter facts. The figure described here is commonly called the figure-eight, although it would be more accurate to call it a double three, thus :—

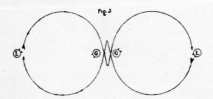

Fig. 3

NOTE.—*In this diagram, and others which will be noted, I am obliged to move the performers slightly from their true positions, in order to avoid obscuring parts of the designs. In Fig. 3 both ladies ought to be placed a little nearer the centre, and exactly upon the track passing before them. Both gentlemen ought to be placed upon the intersections which are almost behind them.*

Leaving the discussion of terms to a more fitting place, let us see what you really have to do.

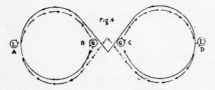

Fig. 4

NOTE.—*See observations on Fig. 3. Note also that the figure or track starting from C, and that starting from B, ought to coincide for the greater part of their path ; for diagrammatic purposes they have been kept separate throughout.*

Suppose you are the dancer B (Fig. 4). During the first eight bars of the tune you will move off to your left with the right foot, and (with the Kemshoole movement) follow the path leading from B, you will finish in C's place and face his partner. During the second eight bars of the tune you will "sett," as it is called, with the 1st Strathspey step. While you are doing your part of the so-called figure-eight, your friend C describes a path the counterpart of yours, but from the point C toward D (see dotted track, Fig. 4), and he finishes in your place facing your partner, to whom he also setts during the second part of the tune. The ladies join in the figure—each from her own point of departure—and they sett while the gentlemen are setting. For a second time the figure-eight is described—you and your friend C starting from each other's place and finishing each in your own—after which you again sett (with the 1st Strathspey step). A third time the figure-eight is described (the gentlemen again exchanging places), and a third time you each sett to the ladies (with the 1st Strathspey step). The music now changes to the Reel. You describe the figure-eight a fourth time, but, instead of the Kemshoole, you must now use the Chassé movement. At the termination of the figure-eight you sett to the lady with the

1st Reel step. The figure-eight (with Chassé), followed by the 1st Reel step, are repeated a second and a third time. Now the Strathspey and Reel or Foursome is ended. It is, however, all repeated from the beginning, and again and again as often as the performers desire.

What I have endeavoured to describe is but the framework. You must perceive how monotonous it would become if you had one step and no more for each division of the dance. It is here that variety shows itself. Before you can call yourself a dancer of the Foursome, or any other form of the Reel, you must acquire two sets of steps—one set for the Strathspey, a second for the Reel. Not only must you have two sets of steps, but you must, by practice, be able to select a step, and perform it with neatness and facility during the whirl of the dance. If each dancer possesses a good stock of steps, and performs them in any order, you will readily conceive there will not only be variety in the work of the individual, but also variety in that of the four performers. Though all in common describe a definite figure or path, and finish in fixed positions, yet each dancer has opportunities, almost infinite, to exercise taste and individuality. Not only is there great variety in the steps used in setting, but, as you will afterwards learn, there are substitutes for the Kemshoole and the Chassé which the dancers may use; therefore, though all move through one path, the movements used by each dancer in describing this path may vary. It is this diversity of detail in the execution of a figure common to all which forms the chief charm of the dance.

FAREWELL.

LETTER IV.

∗ ∗ ∗

My Dear A——

It is my intention in this Letter to describe the form of the Reel of Tulloch.

I take this Reel second because it is frequently danced as a sequel to the Foursome, and is of more practical utility than either the Reel of Three or the Reel of Eight, both of which I will describe in subsequent letters. In my endeavour to give you a clear idea of the construction of this dance I will adopt the method I used in Letter III. with regard to the Foursome. I will describe the movements which are essential to its performance, the position of the dancers at the opening of the dance, and the sequence of parts in the dance as a whole.

Before I describe the essential movements, I should inform you that the rhythm of the music for this dance is that of the Reel—there is no question of the Strathspey rhythm cropping up to embarrass you. There is a well-known tune to which this is usually performed (see end of this letter), but for practice any Reel tune will do.

Essential movements :—

1st. The Chassé.—Already described in Letter III.

2nd. A Reel Step.—For our present purpose we can use the 1st Reel step (see Letter III.)

3rd. Wheeling.—This is a term used to indicate a part of the dance frequently repeated.

It is performed as follows :—Two dancers stand side by side, facing opposite directions, thus :—

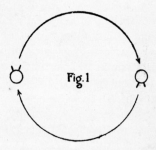

Fig. 1

Each dancer places the left arm across the small of back, and extending the right arm sideward (palm of hand turned backward), passes it through beneath the extended right arm of the other, and grasps his or her left hand with the right. In this interlocked position they each *chassé* round after the other (at arm's length apart) for three bars. At the 4th bar they, quitting hands, make a half-turn (upon their own axis) to their right, and, with arms interlocked anew, but reversed, they chassé to their places in the contrary direction.

You will no doubt be surprised to learn that if you are quite familiar with these three movements (Chassé, 1st Reel step, and Wheeling), you can take part in this dance without inconvenience to the others.

With regard to the dance itself. It is danced by four persons, two ladies and two gentlemen, who place themselves as for the Foursome (see Fig. 2, Letter III.)

Though at first sight the description I am about to make looks most bewildering, yet it is really very simple. If you take four coins or counters (numbered 1, 2, 3, 4), and lay them on a table, you can work out the successive positions of the dancers very clearly.

In order to simplify description, I divide the dance into alphabetical sections.

The dancers place themselves, and proceed as follows :—

① ② ③ ④

(a). 1 and 2, also 3 and 4, sett to each other (with the 1st Reel step), 8 bars, and wheel 8 bars (4 bars in one direction and 4 in the other), after which they finish thus :—

② ① ④ ③

34

(b). 2 and 3 now stand still while 1 and 4 sett 8 bars and wheel 8 bars, after which they finish thus :—

②   ④ ①   ③

(c). 2 and 4, also 1 and 3, sett 8 bars and wheel 8 bars, after which they finish thus :—

④   ② ③   ①

(d). 4 and 1 now stand still while 2 and 3 sett 8 bars and wheel 8 bars, after which they finish thus :—

④   ③ ②   ①

(e). 3 and 4, also 2 and 1, sett 8 bars and wheel 8 bars, after which they finish thus :—

③   ④ ①   ②

(f). 3 and 2 now stand still while 4 and 1 sett 8 bars and wheel 8 bars, after which they finish thus :—

③   ① ④   ②

(g). 3 and 1, also 4 and 2, sett 8 bars and wheel 8 bars, after which they finish thus :—

①   ③ ②   ④

(h). 1 and 4 now stand still while 3 and 2 sett 8 bars and wheel 8 bars, after which they finish thus :—

①   ② ③   ④

which, you will observe, is the arrangement of places from which the dancers commenced.

The dance is now ended.

If you have followed the foregoing description, you must remark the form of the dance is but an alternate setting and wheeling among the four performers. No doubt you will say, "Is this all?" I reply, "No—not all; it is but the framework."

The observations I make in Letter III. regarding the necessity for a stock of steps to give interest and variety to the dance apply as strongly here.

I must not conclude the description of this dance without drawing your attention to two points of considerable importance which you will be expected to know.

1st. In describing the nature of the "Wheeling" movement I said, "with interlocked arms the dancers *chassé* after each other." Yes, the chassé will do very well, but it is not the correct movement for this purpose. Not only is the correct movement difficult to do well, but it is difficult to describe in ordinary language. I will make the endeavour.

Standing (say) upon the right foot—the left being extended in 2nd position—you make a series of, what I can only call, "*slipping* hops" in describing the circular path required in wheeling :—

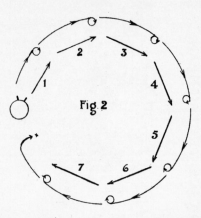

Fig. 2

Each straight arrow, Fig. 2, represents one of these progressive "slipping hops." While these are being executed upon the right foot, the toe of the left foot is made to describe a series of small circles (forward and inward) by the flexion of the knee and the advancing of the whole limb. The outer line, Fig. 2, shows in diagrammatic manner the path of left foot. You will find in practice you require to make seven of these "slipping hops," and at the eighth beat you transfer the weight of the body on to the left foot and return in a similar manner in the opposite direction. This movement is very effective, and should be acquired.

2nd. In explaining the "Wheeling" movement I described the manner of interlocking

REEL OF TULLOCH OR RIGHIL THULAICHEAN.

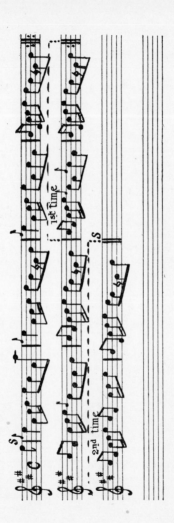

arms. This manner of "linking," though perfectly correct in all cases, is usually reserved for the occasion of wheeling with a lady. When two men wheel each grasps the other underneath the upper arm, and, linked in this manner (with free arms held well up and out), perform the wheeling.

On the preceding page I have inserted the well-known tune for this dance.

FAREWELL.

LETTER V.

* * *

My Dear A——

In this Letter I will endeavour to give you a clear impression of the form of the Reel of Three—not because I think it comes third in importance, but because a knowledge of the *figure* described in this dance is necessary before you can take part in the Reel of Eight, which I intend to describe in my next Letter.

I will adhere to the methods of Letters III. and IV. I will explain :—

1st. The essential movements to be acquired.

2nd. The positions from which the dancers begin.

3rd. The form of the dance as a whole.

1st. *Essential movements.*

With the knowledge you now possess of the Kemshoole, 1st Strathspey step, Chassé, and the 1st Reel step, you can take part in this dance.

2nd. In order to begin, the performers place themselves thus :—

Two ladies and one gentleman, or thus :—

Two gentlemen and one lady.

3rd. *With regard to the dance as a whole.*

Your task is easy. With the exception of a change in the form of *figure* described by the performers, this dance proceeds along precisely the same lines as the "Foursome." It is necessary, before going further, to make this change of figure clear.

In Letter III. I stated that though by common usage the figure described in the Foursome is called the figure-eight, yet the figure described in that dance is not a *true* figure-eight, but is, as I said, more like a double-three. In the Reel of Three, however, the dancers describe a *true* figure-eight, and this I will endeavour to show.

Suppose you stand thus :—

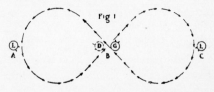

NOTE.—*Both ladies ought to be placed upon the nearest part of the track passing before them—the gentleman should be placed upon the intersection.*

Take B's path as an illustration. He, starting off (with his right foot) toward the left, will follow the stream of arrows, and finish at D (facing A). When, after "setting," he has to repeat the figure, he *reverses* the directions of all the arrows, and returns by the same path. Each dancer follows the stream of arrows from his or her point of departure. If you study the path of each performer you will observe that B and C start off by their *left* (Fig. 1), but A must start off by the *right* in order to avoid collision with C. In practice you will find that A, starting off by the right, will pass *behind* C about the intersection at B.

Before leaving the description of the true figure-eight, I must tell you that the direction in which the dancers describe the figure is a matter of expediency. The dancer B (associated with the others) may, if he chooses, describe the figure thus :—

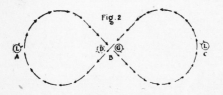

NOTE.—*See observations on Fig. 1 of this Letter.*

You will observe the path described in Fig. 2 is similar to that described in Fig. 1, but the curve is reversed. I would not have mentioned this change were it not that in the Reel of Eight you will find the direction in Fig. 2 more convenient. Do not, however, permit this question of direction to cause you trouble— either direction will do, though that of Fig. 2 is more suitable.

I think, to avoid confusion of terms, we ought to call the figure described in this dance the *True* Figure-eight, to distinguish it from the so-called figure-eight of the Foursome. Whenever, therefore, I have occasion to refer to the figure-eight as described in the Reel of Three, I will call it the *True Figure-eight*.

Having now, I hope satisfactorily, explained the form of figure required in this dance, let us proceed with the dance as a whole.

Your task is again easy. You have but to go back to the description of the Foursome (Letter III.), and, removing one of the centre partners, you carry on the dance as there described—always remembering you now use the True Figure-eight.

If you were to perform this dance several times you would discover the underlying principle to be :—*Two* ladies figure with and sett to *one* gentleman, while he figures with and setts to *both* ladies. "What!" you say, "one gentleman to sett in two directions at the same time? Impossible." Not impossible, but certainly difficult.

With your limited resources the ceremony of setting to *both* ladies would be neither an artistic nor yet a courteous performance, since you would have your back to one lady or the other most of the time you were setting ; but as there are some steps during which the dancer rotates upon his axis, or nearly so, you can readily understand that the use of such steps would enable the performer to give attention to both ladies while ostensibly setting to one.

The position of the centre dancer in this dance is one of a trying nature. It makes large demands upon his resources and presence of mind. He must so skilfully choose his steps that neither lady may feel the pain of neglect, and both the satisfaction of equal attention.

FAREWELL.

LETTER VI.

My Dear A——

If you have made yourself
acquainted with the dances described in Letters
III., IV., and V., you will have little difficulty
in acquiring a clear impression of the form of
the Reel of Eight, or, as it is sometimes called,
the *Eightsome*. This dance, when viewed by
the uninitiated spectator, appears intricate and
involved; to have neither beginning nor end;
a dance in which the different performers seem
to improvise. It requires a quick and trained
eye to detect the recurrence of similar parts.
I need not say this absence of order and parts
is only apparent.

The principle or idea running through the
dance is as follows:—One dancer, going into
the centre of the set, performs a *pas seul*, after
which he or she devotes some attention to each
of the other performers of the opposite sex.
How this idea is worked out I will endeavour
to show.

I cannot do better than divide my remarks as in previous letters :—

    1st. Essential Movements.

    2nd. Position of dancers at opening of dance.

    3rd. The form of the dance as a whole.

Before going further, I may say the rhythm and speed of the music for this dance are the same as for the *Reel*.

    1st. *Essential Movements.*

You still require the Chassé as described in Letter III., and the *Wheeling* movement described in Letter IV. In addition to these movements you will require, in the first place, to obtain a clearer idea of what the Chassé really is, and, in the second, to learn a peculiar setting step which I will call the *Pas-de-basque*, because it is composed of the same elements as that well-known compound movement.

The Chassé. If you go back to Letter III., and again carefully read the description of the Chassé, you observe that one foot is driven before the other *twice*, and so on each foot alternately leading. I have called this the Chassé. Suppose, however, you were *not* to alternate the foot leading, but to drive the same foot continuously with the other in any given direction, this would still be a Chassé movement, but here you have *Chassés desuite*—

meaning, as the words imply, Chassés in a series.

By means of the Chassés desuite you can describe a curved path in a sideward direction, either to the left with the left foot leading, or to the right with the right foot leading. It is this use of the Chassés desuite you will require in the Reel of Eight.

To avoid confusion let us reserve the term *Chassé* for that movement in which each foot leads in turn (as in Letter III.), and the term *Chassés desuite* for the continuous use of the same foot in a given direction.

Pas-de-basque. This is a neat and character-istic movement which usually finds place in this dance :—

Stand on your left foot—the right being held either over the instep of the left or extended in the 2nd position.

a { Spring lightly to the right on to the right foot, immediately placing the left foot behind the right ankle } one
b { Smartly slip the left foot to the left, passing lightly on to it } and
c { As smartly displace the left foot with the right, at the same time extending left foot to 2nd position } two
} 1 bar

Repeat a, b, c, commencing with the left foot to the left, and so on alternately as often as required.

Possessed of these additions (Chassés desuite and Pas-de-basque) to your resources you may

now brave the terrors of the dance, but woe
betide you if for one moment you lose your
presence of mind. I think it was Mark Twain
who, after taking part in a Russian dance for
the first time, said, " The dance was very lively
and complicated. It was complicated enough
without me—with me it was more so."

2nd. *Position of dancers at the opening of the
dance.*

They stand thus :—

1ˢᵗ Couple

3ʳᵈ   Fig. 1   4ᵗʰ

2ⁿᵈ

3rd. *Description of the dance.*

The dance divides itself into three parts :—

I. Introduction.

II. Reel of Eight Proper.

III. Coda.

There is no pause or interval between the
parts.

48

I. *Introduction.*

(a) The eight performers join hands in a large circle, and move round to their places (by their left) with the *Chassés desuite* ... 8 bars

(b) Each lady (still holding her partner's right hand with her left) gives her right hand to the opposite lady. All the dancers now move, like the spokes of a wheel, about half-way round the set (with the *Chassé* movement). At this point the four gentlemen draw their partners outward from the centre and go therein themselves. They (still holding their partner's hands) give their left hands across to each other, and all return to places in a contrary direction 8 bars

(c) At the conclusion of b, each dancer finishes facing his or her partner ; they sett to each other with the *Pas-de-basque* four times and *wheel* (with right arms) ... ... 8 bars

(d) All perform *Grand Chain* as you have it in the 5th figure of the Lancers, but here the Chain is performed with the Chassé movement .. .. 16 bars

II. *Reel of Eight Proper.*

(e) The first lady goes into the centre of the dance and performs her *pas seul* (that is, performs a Reel Step), while the remaining seven join hands and move round to their left, as at "a" in Introduction ... ... 8 bars

(f) The first lady faces her own partner—they do *pas-de-basque twice* and chassé round each other (giving hands—not wheeling) 4 bars

(g) The first lady repeats f with the opposite gentleman .. .. 4 bars

(h) The first lady describes the *True* Figure-eight (Letter V., Fig. 2), with her own partner and the opposite gentleman ... 8 bars

(i) The first lady repeats e, f, g, and h, but now
with the side gentlemen—she then retires
to her original place   ...   .   24 bars

After this the 2nd, 3rd, and 4th ladies, the 1st,
2nd, 3rd, and 4th gentlemen, in their order,
go into the centre and perform e, f, g, h,
and i.

III. *Coda.*

When the last gentleman has finished his
part in the Reel of Eight Proper, the Introduction
is repeated as a Coda—then the Reel of Eight
is ended.

I cannot expect you to follow the foregoing
description without the aid of the other partners
in the actual dance, but you might arrive at a
reasonable idea of the principal parts with the
aid of eight coins or counters, after the manner
suggested for the Reel of Tulloch.

As this letter is already too long, and is full
of matter which requires working out, I will
leave for another letter any further observations
I have to make upon this Reel and upon Reel
dancing in general.

FAREWELL.

LETTER VII.

⁂

My Dear A——

I have reserved for this Letter a number of minor points of interest which, if inserted in their respective Letters, would have diverted your attention from the broad outlines of the dances. I will arrange my observations into four groups :—

1st. The Foursome.

2nd. The Reel of Tulloch.

3rd. The Reel of Three.

4th. The Reel of Eight.

1st. *The Foursome.*

In describing the so-called figure-eight in this dance the performers sometimes omit the winding or sinuous track in the *Reel* or quick time. They describe instead an oval path, thus :—

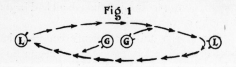

Fig 1

NOTE.—*Both ladies ought to be placed upon the nearest point of the track passing before them.*

I have marked the path for the ladies, but the gentlemen, moving off by their left, join and follow the current of arrows as indicated. Each gentleman finishes by turning sharply into the place of the other.

There is a variation in the opening of this dance which might confuse you were you not acquainted with it. You will observe, from Letter III., Fig. 2, the correct arrangement of the dancers at the beginning of the dance. Sometimes, however, they place themselves thus :—

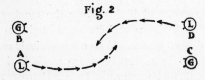

Fig. 2

When this method of opening the dance is adopted, the ladies begin the figure *with* the music; when they have received two bars start, the gentlemen join in the figure, and the dance goes uninterruptedly on as though all had begun from the correct opening.

If you give this a moment's consideration you will discover that if you (B) had begun from the centre, you would have arrived at or about the point B, Fig. 2, in two bars—hence the necessity for the gentleman *waiting* two bars in order to properly adjust the relative positions of the dancers in the figure.

I have already made reference to the fact that it is not necessary to preserve a fixed order of steps in setting. As soon as you have acquired a stock of steps, learn to use them in any order. Be careful, however, you do not attempt to introduce a *Reel* step into the Strathspey, or a *Strathspey* step into the Reel.

Learn as soon as possible to distinguish Reel and Strathspey tunes by their *accent* and *tempo*. The ear is the only guide. If all instrumentalists knew how to play Reels and Strathspeys, and if they knew the number of times each dance and its parts ought to be played, your task would be comparatively easy, but some instrumentalists are weak in both points, therefore you must keep a sharp ear to the music, in order that you may adapt your movements to the music should it, through accident or ignorance, be played without a proper order of parts. Upon no account attempt to dance the steps and movements suited to the one half of the dance to the music for the other.

2nd. *The Reel of Tulloch.*

The only observation I have to make here has been already alluded to in Letter IV., namely :—This dance is often danced as a sequel to the Foursome.

In this case, when the dancers have done justice to the Foursome, the instrumentalist strikes up the well-known tune—whereupon, without interval or pause, the dancers break into the Reel of Tulloch, but (and note this) they begin by *Wheeling.* Once begun, the wheeling and setting go on alternately to the end.

3rd. *Reel of Three.*

I have no observations to make upon the dance itself, but as there happens to be a *Contre-Danse* (commonly called Country-Dance) of the same name, you might be misled.

The Contre-Danse being foreign to the purpose of these Letters, I simply mention this similarity in name. Usually this Contre-Danse is referred to as :—

The Reel of Three C.D. (=Contre-Danse).

4th. *Reel of Eight.*

When this long dance is finished, you would imagine the dancers ought to seek a well-earned rest.   This, however, is not always the case.   In an enthusiastic gathering a curious re-arrangement of places occurs at the end of the dance. I will endeavour to explain this :—When the Eightsome is finished, the instrumentalists make several prolonged strains or chords, during which the 1st and 3rd couples open out and face each other—the gentlemen at the same time going back-to-back between their partners; the 2nd and 4th couples do the same.   The eight performers now stand in two sets, as for the Foursome.   The Strathspey is now struck up, and is performed three times, then the Reel *once*, and in conclusion the Reel of Tulloch is gone through to the end.

Instead of the 1st and 3rd couples, and the 2nd and 4th couples opening out, as above described, the dancers may, by mutual agreement, arrange themselves in any order.

At "a" and "e" in Letter VI. I said the dancers move round the set by the left.   Instead of moving quite round the set in one direction, the dancers sometimes move half-way round by their left and return by their right.   You will require to keep a watchful eye about you and remain in touch with the company.

In the Grand Chain, "d," Letter VI., the dancers sometimes "wheel" their partners with the right arm when they meet half-way round the set, and again on returning to their places.

I must again advise you to be observant, and be prepared to join in any of the small changes you may meet.

FAREWELL.

LETTER VIII.

⊛ ⊛ ⊛

My Dear A——

In the foregoing Letters I have endeavoured to place before you the form and construction of the Reels which receive general recognition. There are, however, other forms which, though they do not receive general recognition, yet possess intrinsic merits sufficient to justify their preservation.

Although the Reels described in this Letter are seldom seen except in the Dancing School, and might, perhaps, have been omitted here, yet I give them in the hope that they may receive a larger share of attention than has hitherto been bestowed upon them. They are beautiful in design, and the skill necessary to their performance is of a high order.

1st. *The Reel of Five.*

In which four ladies and one gentleman stand thus :—

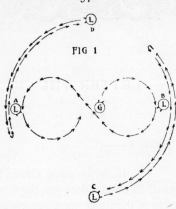

FIG 1

NOTE.—*The ladies A and B ought to be placed upon the nearest point of the track passing before them; the gentleman ought to be placed upon the intersection of the figure-eight. This will enable the ladies C and D to be placed proportionately nearer the centre.*

The construction of this dance is simple. The gentleman, with A and B, dances the Reel of Three. During the figure the ladies C and D chassé (by their right) about half-way round the set and back; they then sett to the gentleman with the others. The gentleman describes the figure with each pair of ladies alternately, and, as in the Reel of Three, he setts to them all by means of rotary steps. I need hardly say that, as each pair of ladies (and the gentleman) describe the figure, the other two describe the curved path outside the centre trio.

58

2nd. *The Reel of Six.*

The dancers stand thus :—

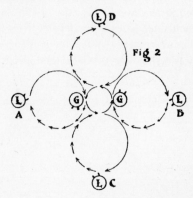

NOTE.—*Each lady ought to be placed upon the nearest point of the track passing before her. The two gentlemen ought to be placed upon the junction of the outer and inner circles.*

Here, again, the construction is simple. The gentlemen, with A and B, begin the Foursome. While the two gentlemen, with A and B, are describing the figure, the ladies C and D move into the centre, and as they approach the other ladies the four move round each other in a wheel-like manner and pass on—each to the further side of the set. They return in the same manner, but, of course, by the reverse half of the track.

You will observe that when the four ladies are in the centre the two gentlemen are at the outside—when the gentlemen, in turn, reach the centre the four ladies are at the outside. The gentlemen figure with and sett to each pair of ladies alternately. When the gentlemen sett to one pair the other two ladies sett also.

3rd. *The Double Foursome.*

The dancers stand thus :—

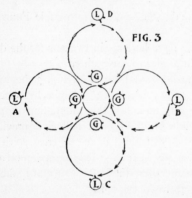

NOTE.—*The four ladies ought to be placed as in Fig. 2 of this Letter ; the gentlemen ought to be placed upon the junction of the outer and inner circles.*

If you inspect Fig. 3 closely you will observe it is simply two sets for the Foursome at right angles to one another. Each set of four performs the Foursome as already described, but when the four ladies approach and meet in the centre they move round each other in a wheel-like manner as in the Reel of Six. The four gentlemen do exactly the same when they meet in the centre. Except for this difference the dance is the same as the Foursome.

This is a particularly pleasing Reel from the spectacular point of view.

4th. *The Double Reel of Tulloch.*

The dancers stand as for double Foursome, Fig. 3.

You have here again two sets for the dance at right angles to each other.

The horizontal set A B, Fig. 3, begins by setting; the vertical set C D begins by wheeling. If the sets maintain this alternation of parts, and follow out the Reel of Tulloch as previously described, you get a dance of great spirit and considerable beauty. This dance receives a well defined termination if *one* extra wheel is given by the four couples.

FAREWELL.

LETTER IX.

My Dear A——

The preceding six Letters have been devoted to the consideration of the various Reels—their form and construction.

It is now my wish to ask your attention to some points of interest and utility applicable to Reel dancing in general. In doing so I wish you to bear in mind that the purpose of these Letters is not to teach you to dance.

What I have written treats of only one small corner of a much larger subject "Dancing," and but scantily even of that. Though making no attempt at instruction in its larger sense, I think I may legitimately lay down some broad rules for your guidance. I will state the broad rules, and rely upon your taste to apply them in the execution of the steps and movements which precede and follow this Letter. At first sight there may seem little or no connection between the rules and the steps described, but if you choose to acquire the steps by rote, even though indifferently executed, you will find the rules I lay down an aid to neatness and correctness of performance in your subsequent practice.

RULES.

I. All the *open* positions (see Letter I.) must be made with a well extended and turned out limb.

II. The toes must be pointed well downward, so that the sole of the foot is not seen.

III. All *close* positions must be made with the feet equally and well turned out. The feet when in the 1st position should contain an angle of about 90 degrees, and this relative position of the feet should be maintained whether they are in a close or an open position.

IV. Whenever one foot is placed close behind or before the supporting leg, the knee of the flexed leg must be laid well out and back. For example—if the right foot is behind the calf of the left leg, the right knee must point directly sideward by opening out the thigh.

V. The body and head must be erectly carried.

VI. When your *right* foot is advanced, your body must receive a slight turn toward the *left*— when the *left* foot is advanced, the body is turned in the opposite direction. Although you turn your body from side to side in performing these movements, you must always present your face toward your partner or the lady to whom you are dancing. This, of course, cannot apply if you perform rotary steps, in which, naturally, you will present your face successively to the partners to whom you are setting.

VII. A pliancy of knee and strength of instep action are absolutely necessary. Nothing is more distressing to an onlooker than the appearance of anyone attempting to dance with stiff knees and flat-footed action. Be sure, therefore, in the first place, you sustain the body by a powerful use of the instep; and, in the second, you modify the many springs and hops by a flexible use of the knees in conjunction with the use of the instep.

VIII. In Reel steps the arms must be either akimbo or hanging by your sides uninfluenced by the movements of the body and feet.

Akimbo has a special meaning here. You place the back of the wrists against the small of your back, keep the thumb and fingers (which ought to be together) extended across the back, and the elbows well forward.

Many dancers—skilled dancers—use their arms overhead and outward in a very effective manner, but this cannot be imparted in a letter—it must be seen to be acquired.

In the Reel steps ladies either place their arms akimbo, or one hand to their dress the other akimbo, or both hands to the dress.

IX. In the Strathspey steps the arms are used in a characteristic manner which I think it possible to state in words. Take for instance the 1st Strathspey step, which I hope you now know. Before you begin the step you have both arms akimbo.

a { During the first four movements of the step you raise the *left* arm overhead and a little in advance of the body.

b { During the second four movements you bring the left arm down akimbo, and at the same time raise the *right* arm as the left was at "a."

c { During the third four movements you bring the right arm down akimbo, and at the same time raise the left as at "a."

d { During the fourth four movements, which are rotary, you raise *both* arms overhead, forward, and outward.

What I have said with regard to the use of the arms in the 1st Strathspey step, applies to most of them. When the *right* foot is in use the *left* arm is raised; when the left foot, you raise the *right* arm; and when the rotary movement occurs, you raise *both* arms.

X. Whenever you raise one or both arms overhead, be careful to avoid *angles* at the elbows and wrists. A line drawn from the shoulder through the arm, and terminating at the finger tips, should be a continuous curve, and not a series of straight lines.

XI. Your fingers must be neatly grouped, and not held out in a fan-like manner. If you make the middle finger and the thumb approach (*without* touching), and do not attempt to restrain the other fingers, they will naturally group themselves round the first two in a becoming manner.

XII. When ladies elect to perform men's steps in the Strathspey, they use their arms as in the Reel steps.

NOTES.

Whenever you have occasion to use the *Chassé* movement, you may substitute a series of light springs from foot to foot (two to each bar) in the direction you require to move. Unless these springs (technically called *jettés*) are made with the greatest neatness it is better not to attempt them, but, if well executed, they are effective, quite characteristic, and give variety to your movements.

Whenever you have occasion to use the *Kemshoole*, you may substitute a series of light springs from foot to foot (two to each bar), each spring being followed by a light hop, during which the disengaged foot is brought to the front. This is also very effective if well performed, but better omitted if not so.

I would advise you to learn a step thoroughly before you attempt the use of the arms.

FAREWELL.

LETTER X.

My Dear A——

In Letter I. I told you I
would divest the dances I wished to describe of
all but the barest essentials. I must confess I
have kept my word to the letter. The steps
and movements described in the preceding
pages are the least you could possibly possess
to enable you to join in the dances without
impeding the other performers.

I have already said the chief charm of Reel
dancing lies in the variety of detail introduced
by the individual dancers.

In the following attempt to add to your
stock of steps, I wish you to understand I can
give those only of easily expressed construction.
It would be impossible without the use of
technical terms to attempt the description of
many which are intricate and beautiful. I
have, however, selected some which are characteristic, well-known, easy of performance, and

which I think it possible to explain without the use of technical terms.

You must not be deterred by the amount of description. When one comes to describe movements in non-technical language it often requires many words to explain one very simple action.

A statesman was once asked to divulge the secret of success in public speaking. He is said to have replied—" The secret is to say nothing in many words." I hope it cannot be said I graduated at this school, though I must confess the insufficiency of ordinary language often compels me to say *little* in many words. I have endeavoured to use the smallest number of words compatible with clearness. If, however, I have used too many, I think I will be forgiven if my descriptions are successful.

You should bear in mind that in dancing— as in many other exercises of skill—he is a better performer who can execute a few simple steps thoroughly well than he who executes many, perhaps more difficult and brilliant steps, indifferently or badly. If you succeed in acquiring those I am about to describe you will be well equipped for the dance, and will not court adverse criticism.

Within the limits of a letter I cannot describe the steps with their many small embellishments—

how to hold and place the feet, legs, body, arms, and head—this requires the example of a living teacher. I can, however, analyse the steps, place their elements before you, and tell you how to re-combine these elements. In addition, I can give some general directions upon the proper disposition of the arms, legs, and feet, and the carriage of the body and head.

In order to learn the steps, as well as for purpose of reference, we will require to number them, but this does not imply a fixed order when used in the dance.

Reel Steps.

First Reel Step. Already described in Letter III.

Second Reel Step. Stand upon the left foot, the right being placed over the left instep.

a	Extend the right foot to 2nd position, at the same time spring lightly on to it.	one
b	Smartly slip the left foot to 4th position in front, and pass lightly on to it.	and	1 bar
c	Smartly bring the right foot in place of the left, which is extended to 4th position in front *raised*	two
d	Smartly bring the left foot down in place of the right, which is extended to 4th position behind *raised*	three	1 bar
e	Smartly bring the right foot underneath and in place of the left, which is extended to 4th position in front *raised*	four

Repeat "a" to "e,' beginning with the left foot to the left, and so on during eight bars of the tune.

This step occupies two bars. In learning it you should count "*one* and *two*" for the first bar, then distinctly "*three, four,*" for the second. This gives a tripping character to the first bar, and a steadier character to the second.

In learning this step as above described, you will find that you cannot avoid advancing each time the step is performed—that by the time you have performed it four times you will be considerably out of your proper place. After a little practice, however, you will discover that by making the *first* movement (that is "a") in a slanting direction backward (not directly sideward) you will be able to retain your proper place however often you may repeat the step. It is advisable, in the first place, to learn steps of this nature as described. After the mechanism is thoroughly grasped, you can then modify as directed, in order to retain your proper place in the dance.

Third Reel Step. Stand upon the left foot, the right being extended in 4th position in front.

a { Hop on the left foot, keeping the right still extended to the front. } one } 1 bar
b { Repeat "a" } two }

c { Bring the right foot smartly down in place of the left, which is extended to the 4th position behind *raised* } three } 1 bar

d { Bring the left foot smartly in place of the right, which is extended to 4th position in front *raised* } four

e { Repeat "c" } five

f { Hop on the right foot, and at the same time carry the left foot to 4th position in front *raised* } six } 1 bar

g { Bring the left foot smartly in place of the right, which is extended to 4th position behind *raised* } seven

h { Bring the right foot smartly in place of the left, which is extended to 4th position in front } eight } 1 bar

Repeat "a" to "h," commencing with left foot. This step occupies four bars.

Fourth Reel Step. Stand on the left foot, the right placed neatly behind and touching the left ankle.

a { Place (do not spring) the right foot down immediately underneath the left heel, at the same time raising the left foot neatly over the right instep } one

b { Hop lightly on the right foot, and smartly carry the left foot closely round and place it immediately behind the right ankle } and

c { Place (do not spring) the left foot down immediately underneath the right heel, at the same time raising the right foot neatly over the left instep. } two } 1 bar

d { Hop lightly on the left foot, and carry the right foot closely round and place it immediately behind the left ankle } and

And so on, foot after foot, during eight bars. When *written*, this step seems commonplace and feeble, but when performed with vivacity, with well pointed toes, and with knees laid well open, it is very effective.

The foregoing Reel steps are suitable for both ladies and gentlemen.

STRATHSPEY STEPS.

First Strathspey Step. Already described in Letter III.

Second Strathspey Step. Stand on the left foot, the right placed closely behind the left ankle.

a { Hop on the left foot, and extend the right to 2nd position	} one	
b { Hop on the left foot, and place the right closely behind the calf of left leg	} two	} 1 bar
c { Hop on the left foot, and extend the right forward to 4th position	} three	
d { Hop on the left foot, and place the right closely before the left leg, a little below the knee	} four	

Repeat "a" to "d" with left foot	1 bar
Again repeat with right foot	1 bar
Then turn upon the right foot to the right as in the 1st Strathspey step	1 bar
Beginning with the left foot, all the foregoing is repeated, the turn will now be upon the left foot to the left	} 4 bars

Third Strathspey Step. Stand as in second step.

a	Hop on the left foot, and extend the right to 2nd position	one	
b	Hop on the left foot, and place the right closely behind the calf of left leg	two	1 bar
c	Hop on the left foot, and extend the right forward to 4th position	three	
d	Hop on the left foot, and place the right neatly a little above the left instep	four	
e	Spring lightly down on the right foot, and immediately raise the left closely behind the right, the toe almost touching the floor, the foot vertically held, the knee laid well open	five	
f	Spring lightly down on the left foot, and immediately raise the right closely before the left, the toe almost touching the floor, the foot vertically held, the knee laid well open	six	1 bar
g	Repeat " e "	seven	
h	Hop on right foot, keeping the left closely behind the right as it was at the termination of " g "	eight	

Repeat " a " to " h " with left foot leading   2 bars

And so on during eight bars.

This is a very effective step. I have described it as though you faced in one direction during its performance. This is not so. During the execution of a, b, c, d, you gradually make a quarter-turn to the *left* when the *right* foot is being used, and to the *right* when you use the *left* foot. At " h " you resume the

position from which you started. These turn
ings, occurring as they do from side to side,
give your body a rotation through 180 degrees.
If you add to this the raising of the *opposite*
arm during a, b, c, d, the raising of *both* arms
during e, f, g, h, and maintain a frontal aspect
of the face throughout, you have a step of considerable beauty.

Fourth Strathspey Step. Stand as for second
step.

a { Hop on the left foot, and extend the right to 2nd position, but here you rotate the extended leg inward at the hip, so that you strike a *false* position—that is, the toe touching the floor, the heel raised, and the outside of the foot presented to the front } one

b { Hop on the left foot, and (still keeping the right leg extended) turn the leg *outward* at the hip, place the heel upon the floor, and point the toe upward } two

c { Hop on the left foot, and bring the right closely before the left ankle, the toe touching the floor, the back of the foot vertically before the left ankle, the knee laid well open } three

d { Hop on the left foot, and place the right heel where the right toe was at " c " } four

} 1 bar

Repeat a, b, c, d, with the left foot ... 1 bar
Again repeat with right foot ... ... 1 bar
Then turn upon the right foot to the right as
in 1st Strathspey step ... ... 1 bar
The whole is repeated, beginning with the
left foot leading ... ... } 4 bars

I need hardly say this is a step with a strong comic element—a step in which you show true and false positions. Do not fear or despise it because it is comic; the step is well-known and quite characteristic. When you add the proper use of the arms, and turn the body, you have a very taking step.

The first, second, and third of these steps are suitable alike for ladies and gentlemen. I must add, however, that when performed by ladies the actions are somewhat modified as regards elevation of the acting foot.

FAREWELL.

LETTER XI.

* * *

My Dear A——

In Letter X. I said the first,
second, and third Strathspey steps, slightly
modified, could be performed by ladies as well
as gentlemen. I meant ladies might, with perfect taste and propriety, do them if they chose;
but there are, in addition to the steps already
mentioned, others of a *terre-à-terre* nature—that
is, steps in which the feet are slipped lightly
over the floor, and in which all high elevations
and violent springs are avoided. Steps of this
nature are usually performed by ladies. I
think, therefore, the description of a few may
be acceptable. I can attempt to describe steps
of simple construction only, and even these,
though typical, will lose much of their character through the clumsy medium of ordinary
language.

Ladies' Strathspey Steps.

First Step. Stand on the left foot, the right extended in the second position.

a { Slip the right foot behind (or before) the left in 5th position	} one	
b { Slip the left foot to the 2nd position, and pass lightly on to it	} two	} 1 bar
c { Again slip the right foot behind (or before) the left in 5th position	} three	
d { Extend the left foot to 2nd position, and at the same time hop on the right foot	} four	

And so on, each foot alternately leading during 8 bars.

This step, like many other neat and effective steps, appears tame and characterless in words, but if the body is well sustained upon the toes, if the limbs are well extended in the open positions, and the feet accurately and equally crossed in the close positions, the step is really brilliant.

The leading foot may be crossed *twice* before or twice behind the other each time the step is executed, or *once* before and once behind. These variations, though simple in expression, give a pleasing variety.

Second Step. Stand on the left foot, the right held neatly over the left instep.

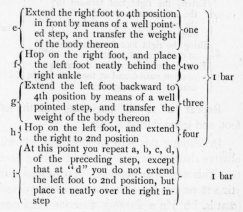

Repeat from "e" to "i," beginning with the left foot, and so on during 8 bars.

———

Third Step. Stand on the left foot, the right extended to 2nd position.

a	Draw the right foot behind the left in 5th position	one	
b	Slip the left foot to 2nd position, and pass on to it	two	1 bar
c	Hop on the left foot, and place the right neatly over the left instep	three	
d	Slide the right foot forward to 4th position, and pass lightly on to it	four	

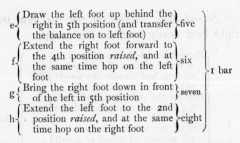

e { Draw the left foot up behind the right in 5th position (and transfer the balance on to left foot) } five

f { Extend the right foot forward to the 4th position *raised*, and at the same time hop on the left foot } six

g { Bring the right foot down in front of the left in 5th position } seven

h { Extend the left foot to the 2nd position *raised*, and at the same time hop on the right foot } eight

} 1 bar

Repeat "a" to "h," beginning with the left foot, and so on during 8 bars.

In learning this step as described you will observe that you cannot avoid advancing each time the step is performed. To obviate this you will require to make the *second* movement (that is, "b") in a slanting direction backward (not directly sideward). See observations on second Reel step (Letter X.)

I feel strongly the foregoing descriptions do scant justice to these beautiful steps. To the actions of the feet must be added the graceful turn of the body in opposition to the advanced foot, the gracious inclination of the head, so that the face is always presented to your partner, the pliancy of knee and powerful use of instep which render the transference of the balance of the body from foot to foot imperceptible, the rounded, extended, and advanced use of the arms, while daintily holding the dress with neatly grouped fingers—these, and other graces, cannot be imparted in words alone, example is necessary.

Those who have not had the advantage of sound tuition in dancing will make but little of my descriptions, but those who have been so fortunate, and who are able to add the graces I have indicated, will find in these descriptions the elements of beautiful steps.

FAREWELL.

LETTER XII.

MY DEAR A——

I will conclude this series of Letters with one or two reflections suggested by their perusal.

I beg you will criticise this little work leniently. It is difficult to describe a simple movement in words, more difficult to describe complex movements, more difficult still to describe one or the other without the use of technical terms. To have thrown the whole subject matter into technical form would have been an easier task for me, but to a reader unacquainted with the technicalities of the art this would then have been a sealed book.

This attempt to describe our National Dances is, as far as I know, the first of its kind. The data for the subject may, in part, be found scattered in books and magazine articles, but, without question, the greater part of our knowledge has been handed down from generation to generation by living teachers, therefore anyone who endeavours to gain a

knowledge of this subject from existing books
will labour in vain.

There being no previous writers on the same
subject, I have been obliged to rely entirely on
myself in the selection of important and the
rejection of less important details, and as to the
form in which the subject is arranged. From
the experience gained in teaching I believe I
have been able to eliminate all non-essentials,
and to present distinct outlines of the dances,
intelligible, when studied carefully, even by those
who are not skilled in the art of dancing.

The diagrams with which the text is illustrated are intended to be suggestive of the
various tracks or paths rather than exact repre-
sentations of the dancers' motions through the
dance. No doubt the dances would be very
effective if performed with the accuracy shown
in the diagrams, but this accuracy cannot be
expected unless all the performers are proficient
and are accustomed to move together.

Reel and Strathspey tunes, when played for
the associated dances, must be played at the
proper speed and with correct accentuation ;
when played as music, apart from the dance,
they may be played according to the taste of
the performer. Some people view the simplicity, and it may be the harmonic transgressions,
of Scottish music with something of contempt,

but these critics should bear in mind that our ancient melodies were constructed on a pentatonic scale—possibly to suit imperfect instruments or the melodic feelings of the composers—and to arrange them in accordance with modern harmonic rules is to rob them of their essence. I have not yet heard of a "Queen's English" edition of Burns, and hope I never may.

It may be said as a reproach that I have treated the subject of these Letters from a standpoint too elementary. I have, it is true, assumed you know nothing about the subject, but, under the circumstances, this was better than assuming, without absolute assurance, you possessed certain knowledge. I place these Letters in your hands, and, whatever be the verdict, I make but one claim—this is, to a sincere attempt to leave a faithful record. My very faults in design and execution may hasten the fulfilment of one of the purposes I have in view in publishing this work—namely, that someone possessing literary experience and a thorough knowledge of the subject may be induced to devote his or her talents to the creation of a work in which our National Dances will receive adequate justice.

FAREWELL.